Astronomy Education Research Primer

Janelle M. Bailey
University of Nevada – Las Vegas

Stephanie J. Slater
University of Wyoming

Timothy F. Slater
University of Wyoming

W. H. Freeman and Company
New York

© 2011 by W. H. Freeman and Company

ISBN-13: 978-1-4292-6409-9
ISBN-10: 1-4292-6409-8

Printed in the United States of America.

First Printing

W. H. Freeman and Company
41 Madison Avenue
New York, NY 10010
Houndmills, Basingstoke RG21 GXS, England

www.whfreeman.com

Table of Contents

Foreword

I t seems that every time we turn around another report is being issued about the problems faced by the United States in the areas of science, technology, engineering, and math (STEM). Organizations such as the National Science Foundation, the U.S. Department of Education, and the National Academy of Science continue to sound the alarm to warn our nation regarding the decline in basic science literacy among our students—our future citizens and workforce. The underlying message is that if the United States is going to continue to be a dominant force in the global economy, and a world leader in addressing the economic and environmental issues faced by the world, then we as a nation need to improve our STEM education practices. In an attempt to address this issue, federal initiatives are implemented and millions of dollars are spent annually, but these resources are limited. If we are to determine the best ways of employing these resources, we must engage in educational research, in a variety of settings, and at multiple scales.

In the summer of 2007, the U.S. Congress passed and President George W. Bush signed into law the *America Creating Opportunities to Meaningfully Promote Excellence in Technology, Education, and Science* Act (America COMPETES). The purpose of this legislation is to advance a comprehensive strategy in a massive attempt to enable the United States to continue to be an innovative nation through strengthening our scientific education and research endeavors, along with improving our technological enterprise and providing job training for our workers to be competitive in the 21st-century economy. While initiatives such as the America COMPETES Act are an important step toward addressing the issues of learning STEM, how do we know if they are successful at improving student achievement and attitudes and making STEM career paths more accessible to the diverse population of our nation? There is clearly an important role for colleges and universities here, but what is it and what information do we need to guide participation of their faculty?

Ask most elementary school children what they are interested in learning about. Two of the most common answers you will hear are dinosaurs and stars. Michael Bennett, the past executive director of the Astronomical Society of the Pacific (ASP) would often quip that these natural inclinations can be the spark to generate a life-long interest in science and technology. We as a nation would benefit greatly by using the natural curiosity about space to encourage our students toward further exploration in astronomy and science in general. As one of the most popular general education distribution requirements among college and university students, astronomy courses can be an important vehicle to build our nation's scientific literacy.

But how are professors and others engaged in astronomy education and outreach to know if astronomy truly is serving as that spark and then making the connection to other sciences? This book candidly and succinctly outlines the impact and value education research provides professors to enhance their teaching of astronomy in order to increase their students learning and understanding of astronomy, as well as science in general.

Not only is astronomy education research itself a valuable endeavor, but so is the process of disseminating and applying the findings. More than 400 years ago, Galileo Galilei's first observations of the stars above forever changed our view of the Universe, but it is through Galileo's writings that we know of his discoveries. As the saying goes, "If it's not in writing, it doesn't count." While this may be the extreme view, it does hold true the importance of documenting your work. With the astronomy education research community growing ever larger, it is valuable to the entire scholarly community to share our research, present our findings, and publish the results. This book, serving as a first-steps "field guide" for new astronomy education researchers, highlights the importance of scholarship within the community and provides pathways to engage in meaningful astronomy education research.

Science and technology issues our citizenry faces are not esoteric areas of irrelevant knowledge off in the distant future, they are in the

4

here and now. Can it truly be that spark of interest in astronomy that is ignited the first time a person looks through a telescope that will generate a greater interest in science and be the first step on the path to improving science literacy within our country? This, and many other wondrous questions, can only be examined effectively through astronomy education research. This book is a great place to start.

Michael G. Gibbs, Ed.D.
Vice President for Advancement
Director, Center for Space Science Education and Public Outreach
Assistant Professor, School of Business and Information Management
Capitol College, Laurel, Maryland

Preface

A mythology exists that most college and university astronomy professors don't really care much about teaching. In this tale, the most senior professors, who are most successful at publishing research, get to teach the small enrollment specialty courses and seminars for the majors or graduate students who most fully appreciate a faculty member's expertise. Furthermore, or so the story goes, the most elite ivory tower professors are rewarded handsomely for winning large research grants by being released from nearly all departmental teaching responsibilities altogether. The losers in this scenario, the junior and less-exalted professors, must then deal with the more challenging, and seemingly less intellectual, large enrollment, general education introductory astronomy survey courses for non-majors. What might seem odd is that this story completely revolves around what professors do or don't do, and not around the students.

To be frank, we don't believe that any of this is true. Not one of us has actually ever heard an astronomy professor say, "If I could just win one more grant, then I'd never have to teach ever again!" (Although, honestly, we have heard our fair share of totally frustrated astronomy professors loudly sounding off when their students weren't achieving at the high level they wish their students would.) What we *do* encounter is one astronomy professor after another who genuinely wants to help their students to grow intellectually and be fundamentally richer human beings as a result of studying astronomy. They want their students to learn more, integrate ideas across domains, see how course topics are relevant to students' lives, develop positive attitudes, and seek out further learning opportunities about astronomy and science beyond the lifespan of a single astronomy course.

Thinking about professors as scholarly and reflective teachers of real students actually makes a lot of sense. In fact, a key feature of scholars and their scholarship is the public sharing of the methods and results so others can benefit from what they've learned. We actually call professors "doctor" because the word "doctor" literally translates from the Latin as "teacher." So, when scholars who are trained in the methods of systematic inquiry turn their attention and honed research skills to issues of education, we find that student learning and attitudes can be greatly enhanced.

At the same time, it has been our experience that simply because someone has been scientifically trained to successfully observe stars or galaxies in order to unlock some of the Universe's secrets, it is not necessarily easy to transfer those same skills and methods directly to studying how people learn. If one knocks a rubber ball off a tabletop to the floor, we pretty much can predict what is going to happen with each and every rubber ball; however, if one knocks a human being off the table and onto the floor, it is nearly impossible to predict what will happen when he or she gets up! Conducting research on human beings is definitely different from conducting research on rocks from Mars. In the words of astronaut George "Pinky" Nelson, "Education isn't rocket science—it's much harder!"

Nonetheless, an astronomer who has abundant knowledge about astronomy, coupled with a strong desire to enhance the student learning experience, is well poised to learn the motivations and methods for engaging in and publishing the results of astronomy education research. This brings us to why we have written this book—we believe that the most efficient pathway to improving the teaching and learning of astronomy is to intellectually engage astronomy professors in rigorously and systematically studying how their students learn, how students interact with astronomical concepts, and how students integrate astronomical ideas into their individual mental landscapes. To help mobilize astronomy professors to develop and begin an agenda in astronomy education research, we present an introduction, a primer if you will, to the underlying ideas of astronomy education research. We by no means intend to provide an exhaustive or definitive work on astronomy education. Rather, we hope to inform and motivate your journey. Along the way, we'll share with you some of the basic concepts

8

and strategies needed to get started, as well as point you to our favorite sources for more detailed information.

Without question, the stated reason for conducting research on the teaching and learning of astronomy is to add to the scholarly knowledge base of astronomy education research. But, it is not our primary intention to turn our readers into education scholars, but to equip them with the tools to improve teaching and learning within their own disciplinary field—if nowhere else but in their own classrooms and departments. The underlying motivation really is about being human— to improve students' learning and life-long affect toward astronomy, and science in general. Welcome to the astronomy education research community!

How to Use This Book

Normally, authors describe how a book is organized and what is contained in each of the relevant chapters in the PREFACE. However, it is our experience that many of our colleagues pass right over the Preface to get quickly to the book's opening pages. We have separated out this guidepost information in an attempt to circumvent such a short-cutting strategy and coax you into reading about how to most effectively use this book. Of course, you are welcome to read it cover to cover starting with page one and heading all the way to the end. Alternatively, if you really just need to find out how to do certain statistical tests of how much your students' learning has changed over the course of a semester, feel free to flip right on over to the QUANTITATIVE METHODS chapter for a summary of how to accomplish this and references to more detailed information published elsewhere.

In the first chapter, our goal is to briefly describe a rationale for conducting astronomy education research. In case this is your first exposure to astronomy education research, we highlight some of the similarities and differences between traditional astronomy research and education research. We also point out why some of your colleagues might initially be somewhat skeptical about your interest in the field and highlight that many scientific professional societies have adopted formal position statements describing the importance of conducting scholarship in the teaching and learning of science.

If you want to know the about the structure of galaxies, computing radioactive decay rates in nucleosynthesis chains won't get you very far. In much the same way, the various questions in astronomy education research call for differing approaches. In Chapter 2, we show you how to match research questions with the most productive research designs that are likely to result in high-quality evidence and publishable results.

In this book, we present proven strategies for attacking problems in the teaching and learning of astronomy. However, the first step is to start with what we already know about teaching and learning, so in Chapter 3 we provide a brief summary that can serve as a departure point in your research efforts. We won't encourage you to test your favorite active learning teaching strategy against the traditional lecture mode—we already know how that study should end up. There is an old saying that applies here that goes something like, "An afternoon spent in the library can save you a month of time in the observatory." Instead, we encourage you to try to uncover the mental mechanisms students bring to the task of learning of astronomy that help us all move forward.

In Chapters 4 and 5, we present two common approaches to collecting and analyzing astronomy education data. On one hand, we introduce some powerful quantitative and statistical tools available. However, just like spectroscopy doesn't provide the answer to every astronomical question, the multiple-choice conceptual test isn't the only tool in the astronomy education researcher's arsenal. While mathematically competent astronomers will probably be most comfortable initially with quantitative approaches, there is considerable explanatory power derived from qualitative approaches. Far from being subjective measures of "Did students like it?" we think you will find qualitative approaches highly attractive because of the deep insights gained about your research question.

As we move toward the end of the book, we focus on the practical aspects of publically sharing and publishing your questions, methods, and results. A key and vital feature of all scholarship is peer review and public dissemination, and the rules of publishing are a little different in both culture and form from traditional astronomy and astrophysics publishing. We point out some of the cultural norms for publishing astronomy education research; these double as pointers on how to create winning grant proposals to support and extend your astronomy education research agenda. Additionally, we talk about some other investigative options that are closely related to research, though slightly different—things like program description and evaluation and action research. The purposes, audiences, and methods of these "other reports" are described in the last chapter.

Insightful education scholars have already spent considerable cognitive energy on a variety of topics, and we see little reason for any of us to reinvent the wheel if it can be avoided. At the very end of the book, we provide a first-steps toolkit appendix that includes some ready-to-go assessment instruments and information on federal regulations about working with human subjects in research (known widely as working with the Institutional Review Board, or IRB). This last appendix, in particular, is important for everyone, regardless of what kind of research involving human beings you eventually decide to pursue.

CHAPTER 1
Introduction: *Why Does a Scientist Need to Read a Book about (Education) Research?*

W ho is the Audience for this Primer?

Through years of schooling, astronomers are trained to have a particular perspective on understanding and representing the way the world works. One way to characterize this scientific worldview is that astronomers use systematically collected evidence and observations in order to form data-based generalizations and inferences about the contents of and physical processes that govern our Universe. Moreover, astronomers who identify themselves as scholars have a moral responsibility—the duty to share what they have learned, both in terms of submitting their ideas to public peer review and also in terms of educating others in our society.

Our goal for this book is to provide a lively introduction for astronomers at any level who would like to know more about how to conduct scientific research in education, particularly in the area of teaching and learning astronomy. Astronomy Education Research, hereafter abbreviated as AER, is our chosen academic scholarly endeavor. We engage in AER at work every day, much like college and university astronomers who are engaging in astronomical research pursuits. We (*try to*) keep up to date with a rapidly growing scholarly literature base, we (*cross our fingers and*) submit grant proposals for resources to make observations and collect data, we visit observing sites (*which are often classrooms or interview studios*), we use observations and experiments to support or refute (*educational*) theories, we teach survey courses for undergraduates (*usually those planning on becoming teachers*), and we supervise graduate students working on Ph.D. dissertations (*most often focusing on vexing problems related to teaching and learning astronomy*). In short, we are writing this book to provide a resource on how to engage in conducting and publishing AER yourself.

Right here at the outset, we need to be clear that this is an astronomy-centric primer for astronomers who desire a "quick start guide" to engaging productively in AER. It is our explicit intention to build awareness, get you started in the right direction, and then point you to our favorite and more complete resources if you need information that is more detailed. Look for the EXIT signs throughout this book as a pointer for where to find more information. For example, if you are looking for a complete reference on how to design, conduct, and evaluate research in education in general, John Creswell from the University of Nebraska— Lincoln has a long track record of publishing useful and quite complete books.

> **EXIT** ▷
>
> See: *Educational Research: Planning, Conducting, and Evaluating Quantitative and Qualitative Research*, 2nd Ed., by John W. Creswell, Pearson Publishing, 2005.

AER's Similarities to and Differences from Astronomy Research

Astronomers are trained to observe objects in the Universe in order to describe what they are, explain how they came about, and predict what will happen to them in the future. Although this is certainly a gross oversimplification, this is done by making systematic observations, developing predictive and explanatory models using mathematics, and looking for connections between observations and theory. In a similar way, astronomy education researchers make systematic observations of learners and their environments in order to develop predictive and explanatory models of education. AER also endeavors to make connections between observation and education theory and, as in astronomy research, the two influence each other greatly. Research articles and meeting presentations look somewhat similar as well (more about this in Chapter 6).

Now, to be fair, part of this similarity is superficial. In recent decades, astronomers have been reporting the results of their education research at professional conferences, such as those of the American Astronomical Society (AAS) and Astronomical Society of the Pacific (ASP), to an audience of astronomers. It's probably not surprising that when astronomy education researchers present their work to people with deep backgrounds in education research and theory, it looks a bit different because the examples, references, and vocabulary can be quite dissimilar, albeit still rigorous. In other words, "When in Rome...". Having said that, one of our hopes with this primer is to begin to bridge some of the gaps between the two worlds.

While there are many similarities between astronomy research and AER, there are also enough differences between the two that one cannot be expected to jump between scholarly endeavors effortlessly. For one thing, galaxies don't have vulnerable feelings that must be respected, or legal rights as human beings do. Moreover, if you have spent time with people, you probably already know that human beings do not necessarily follow any specific and predictive laws of physics in terms of their behavior or learning. (Although most human *bodies*, as opposed to *beings*, do seem to follow the laws of physics when falling.) In the end, the biggest difference is that AER does not have any *a priori* "laws" or "formulae" that predict how all people learn for a given set of initial conditions in the same way we do with physical processes in the natural word.

It is critical to note here that a major aspect of AER that is probably new to you is the issue of "human subjects research." Federal law requires that research performed with human subjects must be approved by an external committee, and you might be required to complete a graded certification test by your institution before you can conduct such research. Important information about the regulations governing and procedures for working with human subjects is described in Appendix C of this book—if you haven't already done so, you'll definitely want to read it.

AER is a Worthy Scholarly Endeavor

AER is conducted in diverse venues. Whereas most academic astronomy is predominately conducted in university physics departments, AER might be done in departments of physics, geology, curriculum & instruction, educational psychology, or cognitive science. The AAS, ASP, American Association of Physics Teachers (AAPT), American Physical Society (APS), American Geophysical Union (AGU), and Geological Society of America (GSA), among others, have strong positions supporting the value of discipline-specific science education research (see the box on the next page as an example). These statements also include support for discipline-based journals in which AER and similar research can be published. For example, the AAS, in partnership with the ASP, sponsors and financially supports a vibrant refereed research journal, *Astronomy Education Review* (found online at http://.aer.aip.org). You can read more about disseminating your work across numerous venues in Chapter 6.

Nevertheless, some academics are openly skeptical, if not hostile, toward the results of education research. (Of course, we should acknowledge that galactic astrophysicists can be skeptical of the value and methods of planetary astronomers' science, and vice versa.) Derision can be easily found within the scientific community and is not solely reserved for the science education field. But to illustrate the status frequently given to AER it is unlikely that any of us would ever stand up in a research colloquium talk about general relativity and insist that some GR data could not possibly be valid because of our anecdotal experience that time seems to accelerate as we mature. Not only would such behavior be considered poor form, but the perpetrator would most likely be censured for asserting an opinion that is entirely "unscientific" in nature. It isn't clear why education research should be treated differently when done in accordance with solid scientific practice, but we have frequently been witness to these types of situations, at colloquia and even at professional conference presentations. The reasons for this are wide ranging, but much of this skepticism and cynicism rests in the following:

Nearly everyone has opinions about schooling, because nearly everyone has been in school. The education community refers to this phenomenon as the "apprenticeship of observation" in each one of us feels as if we have achieved "master" status in education as a result of engaging with education for a number of years. Academics, possibly more so than nearly anyone else in written history, have participated in considerable amounts of schooling and teaching, and have a personal perspective, accurate or not, based on how they themselves think they learned. As a result, some academics believe they are fully qualified to comment on the nature of educational research, even when they are woefully ignorant of the existing literature base or what the standards for evidence are in science education. It is our hope that, as the AER fields continue to mature, and our results are shared at scientific conferences, that our norms of investigation will be increasingly viewed as scientific by the communities that we serve.

American Astronomical Society Resolution In Support of Research in Astronomy Education

In recent years, astronomy education research has begun to emerge as a research area within some astronomy and physics/astronomy departments. This type of research is pursued at several North American universities, it has attracted funding from major governmental agencies, it is both objective and experimental, it is developing publication and dissemination mechanisms, and researchers trained in this area are being recruited by North American colleges and universities. Astronomy education research can and should be subject to the same criteria for evaluation (papers published, grants, etc.) as research in other fields of astronomy. The findings of astronomy education research and the scholarship of teaching, when properly implemented and supported, will improve pedagogical techniques and the evaluation of both teaching and student teaching.

The AAS applauds and supports the acceptance and utilization by astronomy departments of research in astronomy education. The successful adaptation of astronomy education research to improving teaching and learning in astronomy departments requires close contact between astronomy education researchers, education researchers in other disciplines, and teachers who are primarily research scientists. The AAS recognizes that the success and utility of astronomy education research is greatly enhanced when it is centered in an astronomy or physics/astronomy department.

Adopted 2 June 2002, Albuquerque, NM: Available online at URL: http://www.aas.org/governance/resolutions.php#edresearch

At first glance, contemporary astronomy might seem to be *objective* whereas education research is sometimes seen as being *subjective*. Under closer scrutiny, it can be readily argued that if research is done following standard methods agreed upon by the larger community, both astronomy and education research have about the same level of objectivity and subjectivity. In the end, what really seems to be the core issue is that people don't always like the results of education research. We often forget that this happens in a number of *seemingly* objective and value-free scientific domains, too, where people don't like the results and implications of certain research topics. Contemporary environmental science and global warming, astrobiology and the origins of life, and, most closely related to astronomy, cosmology and the Big Bang origin of the universe are just a few examples. Education research, which is clearly value-laden, often falls into the same category as these other areas.

AER is Done for a Variety of Reasons with Results Used in Diverse Ways

People who engage in AER do so for a variety reasons. The most common is probably to improve student learning and attitudes when teaching their own astronomy courses. But, if the efforts and resources poured into conducting an AER study impact only a single astronomer's classroom, we might question if it was really worth the effort. One could argue that most research questions that are only applicable to a single professor's classroom have probably already been solved by the larger science education enterprise. What is most important here is that the real value of AER, when applied to undergraduate astronomy survey courses, is multiplied greatly if the AER effort is focused on a crucial unsolved problem in the larger education domain (this is discussed more deeply in Chapter 2). One example of such an unsolved problem might be the effectiveness of different grading feedback strategies in introductory astronomy courses.

In some settings, a compelling reason to use the methods of AER is to determine the impact of a particular educational program, such as a summer astronomy camp, on its participants. This type of endeavor is more often classified as program evaluation. While an important part of education research, program evaluation has a somewhat different purpose, methodology, and audience. We will discuss this in greater detail in Chapter 7.

The results of AER can be far-reaching. At its most basic level, it informs how professors should design, teach, and assess astronomy courses in order to maximize student learning. It also can impact the nature and design of astronomy texts and web sites used to share the excitement, methods, and results of cutting-edge astronomy. Results that provide insight into how student conceptual understanding evolves provide guidance to K-12 school and science museum curriculum designers. And, not to be forgotten, AER results can help observatory and mission public information officers provide public relations information most efficiently to the scientific journalism corps. The enduring results of AER are those that provide solid information into how learners develop cognitive structures that help them use new scientific information and how attitudes, values, and interests are formed in youth entering a democratic society dominated by information and scientific progress. In the end, we hope that AER helps all of us better understand how to make astronomy education and outreach effective and efficient at all levels.

CHAPTER 2
Research Design: *The QUESTION is the Thing*

R
ecalling the Familiar

Given that our target audience for this book is astronomers and astronomy instructors, who should be well versed in the scientific process, it might seem odd that we are taking the time to walk the reader through the basics of research design. It certainly isn't our intention to patronize our readers, but we have discovered a widespread secret: *Most people, even people with degrees in science, struggle mightily to design research of their own.* It's something that almost no one will admit to (we include ourselves in that), and honestly, it's something we would have rather not found out, but the evidence is too great to ignore.

> **EXIT**
>
> See: *Scientific Research in Education*, Lisa Towne and Richard J. Shavelson, National Academies Press, 2002.

In the past few years we have worked with a large number of people on the topic of scientific inquiry: elementary and high school students, undergraduates, in-service and pre-service teachers, science methods instructors, and arts and science faculty members. With students, our original intention was to provide them with data that they could use to construct scientific ideas. With teachers and faculty, we wanted to provide them with the same data in order to disseminate the curriculum materials more quickly to a wider audience. These interactions took the form of faculty workshops, science methods courses, astronomy content courses, and classroom visits. In every case, with each of these populations, we found that participants struggled to identify, much less generate, fair test questions, and that the people who could craft a coherent research design were few and far between. Surprisingly, people with degrees in science tended to struggle at nearly the same rate as elementary school students, despite their richer catalog of scientific knowledge, familiarity with the research that generated the data, or experience in the lab.

In discussions with faculty participants and science teachers, we found that many of them had never been required to develop research on their own, from scratch. In fact, in most cases the only "science" ever conducted by these folks, who have degrees in science, was done under the supervision of another scientist, and most frequently consisted of data collection and analysis. (At this we sigh at our fond memories of slicing rocks in the basement of the MIT Green Building, digging mud out of the Kansas prairie, or running analyses of spectra, after spectra, after spectra...after spectra... .) We realized, to our amazement, that while we thought we were doing "science" and "research," we were really doing the task of technicians. This is an uncomfortable and rather humbling realization to come to, but the reality is that grappling with a large domain, situating research within that domain, and crafting theory-laden, relevant research, is not common. You may be in the small percentage of folks who have ample experience in this area, but if you have never shaped research from the ground up, you are not alone.

Therefore, before diving into the meat and potatoes of this chapter, the crafting of good research questions, it seems like a good idea to briefly overview research design. This should sound quite familiar to you, as it is exactly the same way one designs research in astronomy, or any other science. This first step is, of course, the development of a research question, a task that sounds deceptively simple. (However, that part is so difficult that we decided to devote the bulk of an entire chapter to the subject.) After spending a great deal of time developing a good question, a researcher turns their attention to the kind of evidence that is needed to attempt to answer the question. They then determine the methodology that is best suited to collecting data. The researcher next determines an analysis protocol based on the questions and available data. When all of these steps have been completed, and you've acquired approval from your Institutional Review Board (see Appendix C at the end of this book), then you can start with the fun of doing research.

> ## Big Idea: Research Design
> *Develop the research question(s), determine the desired evidence, select a methodology to collect relevant data, and determine the analysis technique that will turn the data into evidence that tells us something meaningful.*

We've noticed that many people would like to rearrange this series of steps. In particular, most people would like to design their methodology far before determining what evidence is needed to address their question(s), or they would like to do the two steps together. Also, many would like to delay the discussion of analysis to sometime after they have data in hand. We have some inkling about why this is the case, but that discussion is probably best left for another occasion involving beverages made from hops and barley. The point is that the difficult portion of conducting research usually occurs in the rigorous and intellectually engaging work of designing the work.

The Questions are the Hard Part

Overwhelmingly, the flaw most commonly associated with poor research is a poor research question. This statement probably comes as a surprise, as we like to think that the genius in research manifests itself in clever methodologies or exacting analysis techniques, but that simply isn't the case. The genius in research is best seen in the relevant, insightful, world-altering question. These types of questions are almost always generated by people who know a great deal about the field, who have substantial experience in the field, and who simultaneously have the willingness and ability to look at the field through new lenses. No short order, that. For the rest of us who fall somewhere short of genius, we have to fake brilliance by forcing ourselves to go through steps that mimic what the genius does naturally.

The Lousy Research Question

In the excitement of beginning a new research adventure, it is very tempting to jump straight to the point of the research questions for your study. The problem with doing that is that you are highly likely to develop questions that either: a) are "doable" rather than important, atheoretical rather than theory-laden, or b) don't really address the problem in an informed way.

There are many types of poor research questions, probably including some that we aren't even aware of, but there are at least four forms that we know we all should avoid: "What does this have to do with anything?" "No one else cares about the answer," "But we already knew that" and "Playing with black boxes." Here we provide some specific issues, along with some examples. Since poor research is (usually) vetted out of the system before publication, we will highlight these question forms with examples from the ever-popular science fair. Although these will probably be simplistic and perhaps somewhat pedantic, we hope they will serve to illustrate the idea.

"What does this have to do with anything?" The notion that research should be theory-laden is not one that we talk about very frequently, even though it is the hallmark of good research. Theory-laden research is that which is situated firmly in the context of what is already known about a field, even if it is done in such a way that the researcher is calling the canon of previous understandings into question. Theory-laden research looks at what is known about the field and searches to refine the existing theories or to extend or expand the existing knowledge base related to theory. Alternatively, the research may attempt to question the established point of view by bringing new evidence to bear or by exploring the existing theories by employing previously unused methodologies. Research questions that do nothing to move the field forward, that do nothing to refine the theoretical structure of the field, are wasteful of precious resources. This kind of error is frequently seen in the typical school science fair. For example, consider the problem with the project that investigates the impact of playing music on plants. It might be related to models of plant growth, but we don't know how, and we're betting that none of the students who have done that project do either. To avoid this mistake before engaging in research you have to ask yourself how your project relates to the bigger picture and how it attaches itself to theory. If you cannot answer that question, you need to think about your research questions more deeply before proceeding further.

"No one else cares about the answer." Likewise, avoid research questions that address non-problems in the field of AER. These questions can be answered, but the answers don't move the field forward. These questions tend to either be so personally relevant that they don't apply to the larger community, or they tend to be hair-splitting questions. In the halls of science fairs we see students ask questions like: "Does my dog, Spike, follow commands better when I give him a kibble, or a bit, as a reward?" or "Which cheese molds faster, the cheese from Company A or Company B?" Forget trying to make an argument about the ways in which one could try to generalize from these investigations. Unless you own Spike, or eat cheese from either of those companies, you probably don't care about the answer. Likewise, the larger community is probably not interested in which text works best with a particular institution's population of students, or whether or not a certain group of students liked their course better when lecture slides were placed on a web site. It's not that the answers to these questions have no value. This is the kind of stuff that is great for a discussion on an astronomy teaching e-community discussion group, such as http://groups.yahoo.com/group/astrolrner

These personally relevant questions should be explored systematically, if you think that doing so is going to improve astronomy instruction in your local setting. Action research is the kind of research designed for this purpose. But, action research is not the kind of thing one typically sees reported in the broader literature, because it frequently does not directly add to our theoretical knowledge of how students learn astronomy. (For a discussion of action research, flip to Chapter 7.)

"But we already knew that." Additionally, there are questions for which the answers are foregone conclusions, by which we mean the

answer has already been determined by previous parties. An example from the science fair is the project that asks, "What will happen to my plants if I put them in a very dark closet for a very long time?" Or there is the, "What will happen to my plants if I neglect to water them?" project. We already know the answer to these questions. Some might claim that they don't know the answer to those questions, but we don't buy that argument and believe that such persons should be required to do a little more background research before they are allowed to experiment on defenseless plants. We see some of this same kind of thing in AER and we think it is a result of two factors. First, the majority of people who are interested in astronomy education come from a science background rather than an education background. These folks (including our readers) are experts in the content and methodologies of astronomy, as they should be. They have not had the time, resources, motivation, or extrinsic reward to move them toward becoming knowledgeable in the education literature outside of AER. In this situation, the researcher is operating out of innocence, and can rectify the problem by taking the time to dig into the body of education literature that is out there. Some good places to start for astronomy-specific research are the SABER database (http://astronomy.uwp.edu/saber/) and AER review papers (e.g., Bailey & Slater, 2003, 2005).

In other cases, there are some (who are most likely not reading this book) who do not think that education research has anything to offer them, either because they perceive education research as a woolly business as a whole, or they believe that the larger education community has little that would apply to AER. This is the "there is nothing you can teach me" school of thought. Whatever the cause, asking a question with a foregone conclusion makes the lot of us look silly to education researchers outside of AER. These kinds of questions take the form of "Does active engagement improve learning in the ASTRO 101 course?", "Do formative assessments improve learning in the ASTRO 101 course?", "If I elicit and respond to students' prior knowledge will it improve learning in ASTRO 101 courses?", etc. Well, yes, it will. But answering that question doesn't require a study. There have already been thousands of peer-reviewed and published studies that have established these findings, in every content area, with every age group. (See Chapter 3 for a very brief discussion of some things we already have a pretty good handle on.) The answers to these questions are available in any number of books of teaching and learning. Working on these questions, without further informed refinement, doesn't do much to move AER forward.

"Playing with black boxes." Finally, some questions address a real problem, but reflect a lack of understanding of the nature of the problem. The classic science fair example involves the student who wonders about the effects of common household chemicals on plant growth. The student waters the plant samples with concoctions of soaps and detergents and is moderately surprised to find out that the plant watered with ammonia grows taller and greener than the other plants. Frequently when asked, the student involved in this kind of project can state that ammonia contains a large amount of nitrogen as compared to the other chemical treatments. Unfortunately the student doesn't know enough about plants to know that they tend to enjoy a good dose of nitrogen. The student is conceiving of the chemical treatment as a "black box," by which we mean that the treatment is being studied without consideration for the mechanisms responsible for the outcomes. Reading this you might have felt the urge to smile indulgently at the silliness of the student's naïve scientific work, or you might have shook your head in dismay at the student's lack of theoretical understanding. Either way, you can recognize that the student did a project that was doable, even though they didn't really understand what they were doing. As AERers, we all need to double-check ourselves to avoid doing work of the "black box variety."

The Good Research Questions

When you ask a group of astronomers about the quality of a particular astronomical research question posed during a telescope allocation committee (TAC) review, they all seem to have a gut feeling about what makes a good question without being able to fully define it. It is likely the equivalent of "I know it when I see it," but we don't know anyone that would cop to it in such vague terms. Yet, TAC reviews do seem to do a reasonable job of sorting out research questions that result in fruitful astronomical advances from those that don't.

Judging what makes a great education research question is similar to judging what makes a great astronomy research question—the similarity being that sometimes it's hard to really pin down what makes a great question. Yet, there are some patterns revealing strengths and weaknesses of various questions that we can identify. In our experience, we find that fruitful education research questions have distinguishing characteristics. These characteristics of good research questions usually include, but are certainly not limited to, that they are:

- firmly situated in the literature.
- yield results that move the field forward.
- actionable for the instructor or outreach enthusiasts.
- reveal meaningful underlying mechanisms.

This means that you'll find few enthusiastic readers of your research efforts if you are trying to determine which brand of clickers your students prefer or, in the same way, if jokes and cartoons help students do better on exams. These research efforts just don't sufficiently meet enough of the above criteria.

Beating a Dead Horse

It may seem that we are belaboring the issue of the research question. However, for those who are researchers in the sciences, particularly the field of astronomy where resources are scarce, all of this should make a great deal of sense. After all, there are only so many telescopes and everyone would like to have some time on them. When researchers submit proposals for time on Chandra or Gemini, for instance, there are guidelines for which proposals will receive time or will be placed high up in the batch queue. Highly reviewed proposals will be those that situate themselves within the existing research literature in the field, demonstrate a firm understanding of the theoretical nature of the work, and ask questions that have a high likelihood of moving the field as a whole forward. Proposals that operate at a black box level, that ask questions that the field feels are satisfactorily answered for the time being, or that ask questions that are not interesting or important enough to justify time on a multi-million dollar instrument, are not going to make it.

While no AER project has ever had a price tag to rival these telescopes, the same rules apply. It is tempting to dive into a study that is going to cost little more than instructional materials and time, but just because you *can* do a study doesn't automatically mean you *should* do a study. Research requires that the investigator and others expend time, resources, and mental and emotional energy—resources that could be allocated to other, possibly more fruitful, pursuits. Furthermore, in making the decision to conduct a study, every AER investigator needs to remember that they are not the only stakeholder in the research process and to work that into the equation. For example, if you are using students' class time to explore an educational problem, you need to recognize that the time was not strictly yours to use as you will. You need to be able to make a case that the instructional time being consumed by your study is better used in a study that is *highly likely* to inform the knowledge base than it would be in fulfilling your obligation to provide instruction to the students who have enrolled in (and paid for) your course. So, the bottom line here is, "If you are going to do it, do something worthwhile!"

CHAPTER 3
Big Ideas to Start With: *What Do We Already Know about Teaching and Learning?*

In the long history of the scientific enterprise, we wonder how many times a new, can't-wait-to-get-started, raring-to-go researcher has been gently reminded that, "A long afternoon spent reading in the library can save you wasting a month working in the laboratory." As is certainly true in the history of astronomy research, many creative and intelligent scholars have already allocated considerable effort, if not devoted entire careers, to tackling some of the thorniest issues of the field.

Without question, astronomers would not be too terribly impressed if a professor from the College of Education showed up at an astronomy colloquium lecture and repeatedly interrupted a speaker with uninformed questions that everyone else in the audience already understands. The converse is true as well. Indeed, one needs to know some basics before jumping into any new research field, whether going from education to astronomy, or the other way around. Many of us mistakenly assume that because we went to school, we naturally understand education. This notion is similar to thinking that, because we have played tennis on a sunny day, we already understand solar physics. All silliness aside, the most important idea is that knowing some of the basics allows you to situate your research in the larger context of the discipline and be a meaningful contributor.

In this chapter, we provide you with what we judge to be the five most important ideas uncovered by education research in the last 25 years. This very brief glimpse into what has come before that is not intended to be comprehensive, but rather, to serve as a launching point for your research into the fascinating world of teaching and learning we know as education.

Criteria for Selecting the Big Five Ideas of the Last 25 Years

Selecting and describing the five most important findings in research on teaching and learning in the past 25 years is obviously a subjective task. It is likely that any two people set to the task would select different findings, although one would hope that there might be enough overlap that the two could come to peace over the matter. As we set about determining the most important teaching and learning ideas with you, we first crafted a set of criteria that would serve to both eliminate and recommend inclusions to the list.

The finding had to occur repeatedly in the research literature across many disciplines, in empirical studies and reviews of the literature. Just as in the sciences, education theories need to be supported by evidence, and lots of it. Therefore, every research finding about teaching included here is based on verified research. Groups of independent researchers, including those from related fields such as cognitive psychology, have examined similar conclusions with regard to each instructional intervention. In addition, we believe that findings at this level are only valuable if they have explanatory power. The finding must explain important aspects of what we see in classrooms.

The finding must apply to humans in general, rather than any particular sub-population. We chose to narrow our list to research lines that speak to the way humans learn (and teach), rather than to research specific to segments of the population, or to instructional techniques that are highly specific to a certain content. This criterion serves to exclude some findings that are informative about interactions within the classroom context but that are difficult to translate into instructional strategies. For instance, there has been a large amount of research conducted in the area of gender and education in the past 25 years. From this research we have come to realize that female students and male students frequently perceive and process classroom events quite differently. This is an important finding for teachers to consider as they interact with students, but it is sometimes difficult to prescribe specifically how a teacher would alter instruction based upon this finding alone, and may only apply to half of students. The findings should lead to instructor choices that are likely to increase learning for all students, regardless of context or grade level. A finding should have a sufficiently direct link to the nature of teaching (and learning) in real classrooms that it translates into actionable items for the new instructor at an urban research-extensive university as well as the experienced instructor in a rural kindergarten class. While those two teachers may apply the finding in contextually sensitive ways that appear to be different, they should both be able to apply the finding.

The finding must be pragmatic: not too difficult to apply, with appreciable results. The finding must be pragmatic, translating into a pleasing instructor effort-to-student gain ratio. Teachers who incorporate the finding in their practice should notice improvement in the classroom environment, in terms of learning gains or affective student responses to the learning or classroom environment. These gains should impact a large group of students and be substantial enough to justify the work required for instructors to shift their practices to include the finding.

In the end, we chose five findings that are closely related to the idea that knowledge is socially constructed—that learners alter their ideas and beliefs through experiences and social interactions. This can be viewed as a common theme, not just to our catalog of important findings, but to the field of educational research and psychology in the last 25 years in general.

With that said, from our point of view, research over this time period has yielded firm evidence that the following five teaching principles have the power to significantly improve instruction and student learning and to guide education research.

The Big 5 of the Last 25

1. *Teaching must engage and respond to students' prior knowledge and cognitive structures.*
2. *Learning beyond rote memory requires active engagement.*
3. *Teachers must deliver explicit instruction with regard to the overarching concepts in a content area.*
4. *Teachers must explicitly assist students in developing their metacognitive skills.*
5. *Formative assessments have the power to transform classrooms.*

NUMBER ONE: Teaching must engage and respond to students' prior knowledge and cognitive structures.

Research on students' interpretations of educational events began long ago—at least as far back as work done by Piaget, who was interested in the ways in which young children's cognitive structures supported or hindered their abilities to acquire new knowledge and process information (see for example, Piaget, 1954, 1967). More recent work began with studies in the 70's, such as the work done by Erlwanger (1973), which demonstrated that students have robust alternative conceptions that hide beneath accurately conveyed rote responses. While these two lines of research may not seem immediately related, both investigate students' prior cognitive state: the understandings and mental processes that students bring to the classroom environment, both those that students and teachers are easily aware of and those that are unrevealed.

Prior knowledge, defined broadly, includes students' preexisting attitudes, experiences, and knowledge. The attitudes that learners bring to the classroom encompasses a wide variety of issues: the beliefs about themselves as learners and their conception of their identity, their awareness of their own strengths and weaknesses, and their motivation to

36

engage in classroom processes. Students' prior experiences from everyday activities endow them with a wealth of information to draw from and provide a background for understanding new learning. The work of Norma Gonzalez, Luis Moll, and Cathy Amanti (2005) indicates that everyday family and community experiences follow students right into the classroom and, when acknowledged, can be used as a "fund" upon which new learning can draw, or if ignored, can be a source of miscommunication and conflicting expectations. The knowledge that students have acquired over time includes conceptions of the subject and its content. In some content areas, such as science or history, everyday experiences may have also given students some conception of the structure of the discipline that they are studying: what counts for evidence, and how that evidence is interpreted by those who work in the field.

This area of research has splintered, indeed blossomed, into a wide variety of subtopics, employing a host of associated terms. This work involves prior knowledge, misconceptions, naïve notions, alternative conceptions, phenomenological primitives, and funds of knowledge. Each of these lines hit their stride in the 1980's. In the subsequent 25 years a variety of fields have investigated the ways in which these ideas play out in specific content areas. Findings have provided a means for teachers to improve instruction and increase student learning.

We have learned that just as students can have robust misconceptions (Erlwanger, 1973), students can be assisted in developing robust accurate conceptions (Shepard, 2001). These robust conceptions are in stark contrast to the "fragile" knowledge described by Burns (1993). She convincingly argued that students frequently appear to understand math, and other concepts, but do not appear to understand the same concept when asked to work with this concept in an unfamiliar way. We have come to refer to this as a problem of *transfer*. Burns stated that the implicit classroom expectation is more frequently a matter of students mastering activities rather than concepts. Shepard's response to the pervasiveness of fragile learning is to suggest creating robust learning through "good teaching" which "constantly asks about old understandings in new ways, calls for new applications and draws new connections" (Shepard, 1997, p. 27). In other words, she proposes that the goal is to first elicit students' prior knowledge and, only after that, to assist students in constructing more accurate and more sophisticated understandings.

Preconceptions come into being as natural inferences from everyday experience. If these pre-conceptions are not explic-itly addressed, they will remain as part of many students' mental models, in explicit, latent, or hybrid forms. Instructors must also be prepared to deliberately elicit and think about the mental models that students convey, as students will tend to filter evidence and experience in order to maintain their prior mental models.

> EXIT ▶
>
> See: *How People Learn*, John D. Bransford, Ann L. Brown, and Rodney R. Cocking, National Academies Press, 2000.

NUMBER TWO: Learning beyond rote memory requires active engagement.

The utility of active engagement and social discourse as engaging teaching strategies is certainly not a new finding of the past 25 years. Many of the buzzwords of this period, such as "social cognition," "community of learners," "collaborative learning," "cooperative learning," and "problem-based learning" represent a natural progression of the work began by Lev Vygotsky in the early 1900's and continued by others in the 1960's. "Active engagement" can take on a wide variety of meanings in different venues. This lack of specific definition is appropriate, as there are many strategies by which one may engage students to interact with content, both cognitively and affectively. However, this lack of clarity makes it difficult to clearly define the field of study and complete reviews of relevant literature. In addition, the breadth of the term "active learning" has caused some to suggest that the term itself is meaningless. After all, some might ask, isn't all learning fundamentally active? Is there such a thing a passive learning? For the purpose of this section we have chosen to define active engagement as any instructional method that engages students in the learning process, in which the student must participate in meaningful learning activities and think about what they are doing. While this is a fairly vague definition, it should stand in stark contrast to much of traditional instruction which involves lecture, teacher-talk, and assignments that are rote in nature.

> **EXIT**
>
> See: *Using Active Learning in College Classes: A Range of Options for Faculty*, Tracey E. Sutherland and Charles C. Bonwell, Jossey-Bass, Inc., 1996.

Bonwell and Eison (1991) summarize the literature on active learning and conclude that it leads to better student attitudes and improvements in students' thinking and writing. They also cite evidence from McKeachie (1972) that discussion, or social discourse, surpasses traditional lectures in facilitating retention of material, motivating students for further study, and developing thinking skills. Felder and his colleagues (2000) include active learning in their recommendations of teaching methods that work, and Chickering and Gamson (1987) include active engagement as one of their "Seven Principles for Good Practice."

Seven Principles for Good Practice in Undergraduate Education

1. *Good Practice Encourages Student-Instructor Contact*
2. *Good Practice Encourages Cooperation Among Students*
3. *Good Practice Encourages Active Learning*
4. *Good Practice Gives Prompt Feedback*
5. *Good Practice Emphasizes Time on Task*
6. *Good Practice Communicates High Expectations*
7. *Good Practice Respects Diverse Talents and Ways of Learning*

Adapted from Chickering and Gamson (1987)

In an exhaustive review of the literature on collaborative learning, Johnson, Johnson, and Smith (1998) found that 90 years of research indicated that collaborative work improves learning outcomes. To add even further evidence, a review completed by Springer and colleagues (1999) found similar results by looking at 37 studies in science, technology, engineering, and mathematics. These gains include content knowledge, student self-esteem, attitudes, and students' perceptions of the social support in the classroom. These reviews looked at the size of the gains relative to each outcome, disaggregated to look at "collaborative versus individualistic" studies and "collaborative versus competitive" studies. In both cases, the sizes of the gains were significant, with the largest impacts found in studies comparing the positive effects of collaborative treatments to far less effective competitive treatments. In addition, they found that the biggest impacts were seen when instructors employed a moderate, rather than large, amount of collaborative work in their instruction. Their review also indicated that collaborative work enhanced retention in STEM programs, particularly for students from traditionally unrepresented groups.

There have been a vast number of studies conducted on active engagement. For example, physics education researchers have shown that there are measureable differences in the conceptual models of motion between students who received traditional, largely lecture-based instruction in physics, and those who were in courses that used active engagement as an instructional strategy. In these studies, active engagement was defined as nearly anything that was not lecture, including "clicker" questions, labs, and think-pair-share questions (Hake, 1997; Laws, Thornton, & Sokoloff 1999; Redish et al., 1997). The studies conducted by Redish and his colleagues at Maryland, and Laws and her colleagues at Davidson, suggested that the impact of active engagement was not due to increased time on task, but to the nature of the tasks themselves. Furthermore, student engagement was particularly effective in addressing student misconceptions in physics (Redish et al., 1997). Francis, Adams, and Noonan (1998) found that the difference between control and treatment groups was still measurable one to three years later and that there was only a scant decay in achievement over that time.

Perhaps what could be most surprising is that very small efforts toward actively engaging students can reap appreciable results. In 1987, Ruhl, Hughes, and Schloss found that simply pausing during a lecture to allow students to consolidate their notes, three times, for two minutes at each pause, increases student learning. These researchers found that when an instructor paused for longer periods of time, ranging between 12 and 18 minutes, and allowed students to rewrite notes or discuss the material with peers, without any interaction with the instructor, student learning increased, as measured immediately after instruction and on assessments 12 days after the last lecture. Other interventions might be as simple as allowing students to think or talk about what they know about a topic before formal instruction begins.

In a variety of contexts, with a wide variety of participants, studies have repeatedly shown that actively engaging students in the material increases their learning. The details of why and how this occurs may not be as important for the instructor as it is for the educational researcher, but this is one of the most useful results for increasing students' learning. If instructors employ these teaching strategies with their students, learning will be much more likely to happen.

NUMBER THREE: Teachers must deliver explicit instruction with regard to the overarching concepts in a content area.

Explicit instruction is not the same as direct instruction. Direct instruction is best characterized as the professor telling the students the information and expecting rote memorization in response. Explicit instruction, on the other hand, is best seen when teachers deliberately lead students' attention to the connection between facts and big theories, or when they guide students to notice when the activities of learning mirror the processes inherent to the structure of that particular discipline. This is in contrast to implicit instruction, which rests on the assumption that participation in academic activities will result in an understanding of content and concepts, including those that might be considered overarching.

For example, Brown and Kane (1988) found that generalizable, flexible, and transferable learning is more likely to occur when teachers point students to general themes or to features that suggest solution strategies. In studies of pre-service teachers, researchers have found that implicit instruction of the nature of science and inquiry is not effective. By contrast, science methods courses that deliberately combined content and process, and explicitly engaged students in distinguishing between the two, are effective in increasing pre-service teachers' conceptions of the nature of the scientific enterprise (Gess-Newsome, 2002).

NUMBER FOUR: Instructors must explicitly assist students in developing their metacognitive skills.

Metacognition refers to a person's knowledge of themselves as an information processor. In the classroom this frequently reduces to methods in which learners become aware of and exert control over their own learning. Early work on metacognition was conducted in laboratory settings with young children, such as work reported in the mid-70's, in which Brown (1975) found that rehearsing a memory task with young children resulted in them being able to recall pictures almost as well as adults. Later research extended the work to examine the results of teachers explicitly instructing students in methods of monitoring their own learning.

We have learned that skillful and effective learners are those who take charge and responsibility of their own learning using a variety of "self-check methods" (Brown, 1994) or "executive processes" (Sternberg, 1983). We have also learned that these abilities can be taught through socially mediated processes (classroom activities), just like many other skills (Shepard, 2001). Palinscar and Brown (1984) found that students' reading comprehension improved when they were taught strategies such as questioning, clarifying, and summarizing. They found that these skills, along with predicting, improved students' reading comprehension as compared to a non-treatment group of similar students. In their study, social engagement between students was an additional variable that, they concluded, functioned to reinforce skill-building. When Rosenshine and Meister (1994) used this same technique with learning disabled students, they found that reading scores increased dramatically.

NUMBER FIVE: Formative assessments have the power to transform classrooms.

Traditionally, assessment has been summative and evaluative in nature, in which data are collected from students at the end of a topic or course and assigned a value in terms of a course grade. These values may be archived in databases for accountability purposes, where they function as evidence in judgments made with regard to schools or teachers. At times this same data may be used to assess individual students. In that case, the data may be used to screen a student for special services, graduation or advancement, or to assign grades of a cumulative nature. These types of assessment are by and large separate from the learning process and are similar to the course evaluations that our students submit at the end of the term. Most instructors would assert that these evaluations are a means to allow students to grade us as instructors, and will serve as a tool for administrators who must judge us.

> **EXIT**
>
> See: *Classroom Assessment Techniques: A Handbook for College Teachers* (2nd ed.), Thomas A. Angelo and K. Patricia Cross, Jossey-Bass, Inc., 1993.

In contrast, formative assessments are meant to provide students with feedback in order to assist them in monitoring, and adjusting, their own learning practices *and* they serve to provide the instructor with information that they can use to inform and adjust instruction: a win-win for the student-teacher relationship.

According to Shepard (2001), formative assessments support teaching and learning from a constructivist, or social constructivist, perspective, as opposed to the previous models of assessment which supported older paradigms of teaching and learning (e.g., hereditarianism, social efficiency, and scientific measurement). In this sense, formative assessments can substantively improve classroom environments and enhance learning through positive feedback. What happens is, as formative assessments are adopted, the social interaction between the teacher and the students, and between students, changes in important and not-so-subtle ways, altering the classroom environment for the better. Stipek (1996) found that as the classroom environment changes in this way, motivation is enhanced for all participating students. Moreover, as student motivation increases, active participation in formative assessment increases, causing a positive and much-appreciated feedback loop. In addition, an inference drawn in the study is that these assessments serve to make errors, mistakes, and incomplete understandings an everyday part of learning.

As teachers use formative assessments to shift away from feedback and evaluations that focus on how students perform relative to one another and toward actionable feedback that pinpoints specific learning goals that need improvement, teachers naturally restructure the relationship between teacher and students in order to counteract negative pupil habits (Perrenoud, 1991) and teachers' own habits (Tobin & Ulerich, 1989). In the end, as teachers shift their perception of the function of assessment, their metaphors for their own roles in the teaching, testing, and grading process shift as well, from that of "fair judge" to one of a practitioner trying to create interactions and teachable moments with a window into students' minds.

Pre-instructional, formative assessments can be very useful in attempting to elicit prior knowledge; however, they should be written in the natural language of the learner, taking different contexts into account (Heath, 1987). Otherwise, some students may appear to be deficient because they don't speak school-ese very well. Others, including Minstrell (1989) and Yackel, Cobb, and Wood (1991), also concluded that prior knowledge checks, when manifested in objective test format, are prone to underestimate the knowledge of all but the most sophisticated students. This is most likely because the most traditional school testing formats do not encourage students to translate pre-existing knowledge from other contexts. These researchers suggest that open-ended probing is more likely to elicit coherent and useful explanations of students' initial understandings. This, in the end, guides a teacher's instructional approach that can maximize student learning.

CHAPTER 4
Quantitative Methods: *Show Me the Numbers*

A stronomy by its very nature is a quantitative science. We count stars in clusters, determine planetary diameters, measure the galactic distances, quantify the chemical compositions, and even calculate the age of the universe. So, one might assume that AER is also predominantly a quantitative science. As it turns out, AER is characterized by a healthy dose of both quantitative *and* qualitative measures and both have an important role in AER's scholarly progress. As a place to start, this chapter will deal with quantitative methods of conducting AER and we'll hold off a discussion of qualitative methods until next chapter.

For mathematically trained astronomers, the statistics we'll present here are relatively straightforward, yet will serve the emerging AERer well. The numerical recipes used in measuring how people learn, called psychometrics, can get complex quite quickly and great references for working with these are found elsewhere. But, just because some of these statistical analysis methods are somewhat easy to apply, it doesn't mean that they aren't powerful tools that require great care in using them as part of a larger analysis. Use these critically and they will serve you well.

Analyzing Multiple-Choice Content Knowledge Test Scores

A first step toward measuring students' knowledge is to give them a test. Although you could conceivably create a test on your own, you can probably guess that doing it right takes considerable effort and expertise. In fact, entire AER Ph.D. dissertations taking years of concerted effort are written on the development and validation of diagnostic tests for use in research, much like astronomers take years carefully crafting instruments for use with telescopes. Fortunately, there are numerous reliable and valid diagnostic tests available to use. As but one example among many available, the *Test Of Astronomy STandards*, or TOAST, is provided for you in Appendix A at the end of this book.

What software do I need? There are numerous test scoring and statistical software packages available for use by AERers. It is not our intention to endorse any one particular package or approach, but we find that Microsoft *Excel*, or any of the *Excel*-lookalikes, are perfectly adequate for many scenarios and you probably already have it on your computer. If you must spend your hard-earned money on a software package right this instant, *SPSS* is a popular one, as is *ReMARK*, but these are probably overkill and the learning curve is not insignificant. Also, many colleges and universities do have site licenses for statistical software packages.

How do I get the results into my computer? The most basic way is to simply hand enter it into a spreadsheet. Most people use a single horizontal row for each student and vertical columns for student responses and scores. When we work with lots of students, we will often have students complete scannable bubble sheets, which are then automatically scanned by an optical mark reader (OMR) and deposited into a spreadsheet. Don't completely discard student identification, as you might need it for more detailed analysis when comparing results across different measurements.

What do I get to calculate first? As an astronomer, you probably can't wait to get numbers out of your data! But before you do, you need to "clean" your data, just like you would need to remove cosmic ray hits or subtract the sky from a CCD data set. Take a long look at student answer sheets and develop criteria for which sheets you are going to discard. We suggest that the data from any student that marked answer "A" for every question might be worth discarding. Another one is where students have created a non-random, yet aesthetically pleasing, pattern of responses, such as "A, B, C, D, A, B, C, D, A, B, C, D…". You might also discard sheets where less than 80% of the items are completed, or if the last 20% of the answers are all the same response, or no response, indicating that students may have rushed to finish the task. Only after cleaning the data can you calculate something, most likely starting with the most common descriptive statistics: the number of scores (*n*), the mean (*x*), and standard deviation (*SD*).

So what do I do with these numbers? As you probably already know, the mean (*x*) tells you something about the central tendency of the data, whereas the standard deviation (*SD*) indicates how much variance there is in scores among all the students. To help you have a feel for what this means, 67% of scores are within one standard deviation about the mean—33% of the scores are above the mean and 33% are below. A large standard deviation is interpreted to mean that scores are all over the phase space and it makes it difficult to make generalizations about a particular group of students. When reporting statistics, it is important to always include the number of students in the analysis.

> **EXIT**
>
> See: *Statistics: A Spectator Sport* (2nd ed.), Richard Jaeger, Sage Publications, 1990.

How internally con-sistent and reliable are my scores? Computers are notoriously good at making lots of calculations seem effortless, so you might as well take advantage of it. A common statistic designed to consider if students are answering items consistently throughout the test, and not just getting items right at the beginning and missing all of those at the end, is the *Cronbach alpha*. The *Cronbach alpha* statistic is actually a measure of inter-item correlations (with a value ranging between 0 and 1), and can be interpreted as the extent to which an instrument is internally consistent. A low score should set off a variety of red warning flags that indicate problems like students exhibiting test fatigue or the presence of some unexpected underlying factor, such as the test being more of a measure of a student's reading ability than his or her conceptual understanding. A widely accepted social science cut-off is 0.70 or higher for a set of items to be considered internally consistent.

How difficult are my test items? The degree to which a particular item is reliably assessing students is actually quantifiable in a process known as *item analysis*. If you are using a high-power software analysis tool—or in some cases, the software that comes with your OMR scanner—this can be done automatically. One computation is *item difficulty*. This is a number between 0 and 1 that describes the proportion of students who answered the item correctly. You may notice that this statistic should probably be renamed item *easy-ness* because a high value means that most students answered correctly. For example, when an item has an item difficulty of 0.30 it means that only 30% of students answered this item correctly. Depending on the exact scenario, this would probably be a difficult question, a poorly constructed question, or one that is probing a common misconception. In our experience, it could also mean that you've made an error in creating your test key!

Do my test items discriminate students that get it from those that don't? A second statistical test commonly computed for each item is known as *item discrimination* or *biserial-R*. The item discrimination is a value ranging from −1.0 to +1.0 (although more typically it ranges from 0 to 0.5) and is essentially a correlation statistic. The statistic compares student responses on that item to how students score on the exam as a whole. A high item discrimination value (often 0.3 or greater) suggests that students who scored high on this exam overall tended to get this particular question correct, while students who scored low on this exam tended to get this particular question wrong. In other words, answering the question correctly is correlated with overall mastery of the material. An item discrimination score near 0 indicates that answering the question correctly has nearly no relationship to overall mastery of the material, while a negative biserial suggests that students who generally scored high on the exam fared poorly on this item, while low-scoring students, who otherwise demonstrated a lack of mastery of the material, did well on this item.

In general, a single test covers material within a single domain, although any test might have questions that represent a subsection, or "subscale" within that domain. One would expect that students who have a good overall mastery of the material on the test would also be more likely to answer any given question correctly. When that is not the case, there is cause to examine the test item. Item discrimination values near zero usually mean that students are simply guessing and the item doesn't successfully discriminate between students who generally know the material and those who don't. This may be the case when the item tests a highly robust nonscientific misconception, for instance, an item that probes the test-takers conception of just exactly where the mass comes from in a two-ton tree. Most people, even very intelligent, well-educated people, think that it comes from the soil, or nutrients in the soil. It doesn't. An item testing this content would be likely to have a near-zero point biserial, especially on a pre-test. If such a robust alternative conception is not being tested, items that have low or negative discrimination should be significantly revised or even discounted in calculating students' scores (assuming that this does not indicate an error on the answer key). One word of caution is in order here: If a very high percentage of students get a particular item correct, it will inevitably have low discrimination because the strong students did no better than the weak ones. This does not automatically mean that a particular question is flawed. It is possible, especially in the case of a post-test score, that students have simply attained mastery of the concept in question. If the concept is worthy of student mastery, this is a situation to celebrate!

Comparing Students' Test Scores Before and After an Educational Intervention

A common research design AERers employ is to survey students' understanding at the beginning of an educational intervention, and then again afterwards, to look for changes in student scores. This approach is sometimes known as a ***pre-test, post-test research design***. It is based on the notion, called *process-product*, that if one applies an educational *process* to learners, the end *product* of enhanced learning achievement or attitudes can be measured and compared to the initial conditions. The process, or intervention, here could be a particular lab exercise, inquiry-based unit, or even the whole course—it just depends on what you're looking to test.

Calculating GAIN scores. If student diagnostic test scores are collected at the beginning of an intervention and again at the end, the gain for a student is simply calculated as the difference between the post-test and pre-test scores:

$$Student\ gain = post - pre.$$

If we are talking about the entire study population:

$$Population\ gain = x_{post} - x_{pre}$$

where x_{post} and x_{pre} are the averaged scores for the sample, post-test, and pre-test, respectively. This calculation does give one an important sense of how participants, as an overall group, have changed over time, but it is the least powerful of the statistical tools available.

A more robust strategy for determining students' changes, using a pre-test–post-test design, is to *match* individual students' pre-test scores to their individual post-test scores and determine each student's individualized gains. Those individualized gains are then averaged for the entire class:

$$Averaged\ student\ gain = \frac{sum\ of\ students'\ individual\ gains}{number\ of\ students}$$

There are a couple of tricks that make this a bit more challenging, the first being that the researcher must keep some sort of identifying information with each student's score over the duration of the study. (Whether this identifying information will be easily known, such as the student's name, or some anonymous but unique code, will depend upon your research design and the process required by your IRB.) Further, any student data that lacks a pre- or post-match is automatically removed from the study. This removes the data for any student who began the course late, or who dropped the course. While this makes the sample size smaller, and therefore makes it more challenging to present that highly desired "statistically significant" result, we think that most would agree, students who did not provide data on either end of the study muddy the waters. It may seem that, arithmetically, there is little difference between the two averaging methods, but there are the issues of "clean data." Moreover, when engaging in more sophisticated statistical methods, matched data is always preferred to unmatched data. There is a difference.

Calculating Normalized Student GAIN scores. Because pre-test scores can impact the possible achievable gain, a common strategy is to "normalize" the scores. Normalizing scores is a way to quantify:

$$\frac{\{\text{How much a student has learned}\}}{\{\text{How much the student could have learned}\}}$$

We have used this technique in studies where we were interested in conceptual gain, rather than the final raw score. This formula effectively removes the data for students who arrived in class with a complete conceptual understanding of the material, and allows us to consider the impact of instruction for those students who had something to learn.

Normalized student gain, $<g>$, is calculated by dividing the post-test to pre-test gain by the deficit score from the pre-test, all in percentage form, for each individual student score:

Normalized student gain = (*post*% - *pre*%) ÷ (100% – *pre*%)

In order to calculate the mean for the group, simply average the results for all students:

$$\text{Averaged normalized gain} = \frac{\text{Sum of all students' normalized gains}}{\text{number of students}}$$

Note that students who had perfect scores on the pre-test will drop out of the calculation. (The denominator will go to zero, rendering that data meaningless.) Also, this calculation can only be accomplished if you possess data that is matched, by identifying information, on pre- and post-test. That's not easy to come by, but if you have it, it provides for very robust analysis.

When lacking that kind of matched data, many resort to an average normalized gain. This statistic calculates the average gain of the study group in relationship to the gain that the group could have achieved:

<u>{How much the class improved}</u>
{How much the class could have improved}

The statistic would be calculated like this:

$$\text{Population's normalized gain} = (x_{post}\% - x_{pre}\%) \div (100\% - x_{pre}\%)$$

where "$x_{post}\%$" is the class average on the post-test and "x_{pre}" is the class average on the post-test. This approach is widely known as the normalized gain, or Hake gain, after an oft-cited work by Richard Hake (1998). Hake widely advocated this approach during his 6000-student survey of *Force Concept Inventory* (FCI) scores when demonstrating the measurable impact of interactive-teaching methods over traditional lecture-based teaching methods.

In general, the statistic will give you a ballpark on what is going on, but there can be complications. The researcher has to make choices: Will you include data for students that did not take either the pre-test or the post-test? Will you include data for students that showed a *loss* on the post-test? How will you account for students that received a perfect score on the pre-test? Each of these decisions has implications and must be taken into account when interpreting the statistic, and should be reported in detail in your study report. Despite its far-reaching use, this approach is not without criticism and should be used with some caveats (*c.f.* Brogt, Sabers, Prather, Deming, Hufnagel, & Slater, 2007).

Comparing Mean Scores between Groups. An approach that increases the sophistication of research to the level of a more experimental-like design is to compare the results of two different groups of students when they have received different treatments. This is commonly known as a **multiple-group, pre-test post-test research design** and is a common AER strategy. New AERers sometimes mistakenly describe this approach as a study with a control group, but unless one of the participant groups get no treatment whatsoever, such as comparing a *French 101* class to an *ASTRO 101* class, this approach is more accurately thought of as a *treatment-1* vs. *treatment-2 group* study. Like before, the strategy here is to survey students' understanding or attitudes at the beginning of an educational intervention and again afterwards to look for changes in student scores and then compare the gains.

A Highly Desirable Study Design: Two-Group Pre-test-Post-test Comparison

	Pre-test Score	Post-test Score
Treatment A	A_{pre}	A_{post}
Treatment B	B_{pre}	B_{post}

Whereas simply reporting average scores is known as *descriptive statistics*, using statistics to draw conclusions that extend beyond the immediate data alone is known as *inferential statistics*. The most common inferential statistic used in AER is to compare means by use of the *t* test. The *t* test of means is performed easily with *Excel* or with *SPSS* and yields a *p*-value. To oversimplify this nearly to the point of being criminal, the *p*-value describes the percentage chance that two averages are the same. It's a probability. The calculation assumes that the distribution of scores are somewhat normally distributed, and takes into account the spread of the data in terms of means and of the standard deviation. In general, the larger the number of students, the easier it is to distinguish between the two averages, and therefore, larger sample sizes are more apt to result in a statistically significant result. This test will work for relatively small numbers of students, usually 20 or more.

When performing the *t* test on data, the user typically has two options: You can use a paired-sample *t* test if you have matched data, e.g., pre-test to post-test for individuals; otherwise, use an unpaired, independent *t* test. If given the choice, we'd rather see a matched data *t* test.

What most of us hope for, when comparing post-test to pre-test or *treatment-A* to *treatment-B*, is that there is a statistically significant difference in mean scores. This is best shown when the *p*-value (the probability that these two sets of data are the same) ends up being rather small, at a minimum, smaller than 0.1. Two words of caution here: First, if in your study, you decide that you are going to call any *p*-value less than 0.05 "statistically significant," then any *p*-value less than 0.05 is "statistically significant." By this we mean that, statistically, by the rules that we have set up ourselves, we have determined that the two data sets are, in fact, different. If one *t* test in the study   is 0.04 and another *p*-value is 0.01, one is not *more* statistically significant than the other; rather, we say they are both statistically significant. Being "statistically significant" is, as the saying goes, like being pregnant; there is no "almost statistically significant," and no "very statistically significant." You either are, or you aren't. It's a binary state.

Second, the *p*-value required for statistical significance is not determined *after* the calculation is made. It's determined by the nature of the experiment, which is a subtle topic that we can't really delve into here, and by the culture of the field. In educational research a *p*-value of 0.05 or lower is typically selected. So, it's poor form to say that a *p*-value of 0.05 is your cut-off, and then decide to change that choice after the fact. If you choose 0.05 and one of your results gives you a 0.10....well, that's the way it goes. It's cheating to decide that you'll now call a *p*-value of 0.10 "statistically significant."

Finally, just because the difference in two means is statistically significant, that doesn't mean that it is important. For example, if somehow a pre-test–to–post-test change of 51% to 53% correct turned out to be statistically significant, it wouldn't mean that it was important. If you turn your entire classroom inside out for an improvement from 51% to 53%, you might decide that a 2% increase is not worth the grief, even if it is "statistically significant." This reason, among others, is why we need to remember that this is called "inferential statistics." That means that one must make judgments, and that expertise, whether of a phenomenon, or of a local setting, is required to make meaning of the numbers.

Another approach some people find useful, but beyond the scope of our concise description here, is to calculate an *effect size* statistic, which is the difference in the means divided by the standard deviation of the comparison group or pre-test score. An easy reference, including an *Excel* worksheet, for understanding and calculating effect size is provided by Thalheimer and Cook (2002).

It is further worth noting here that, many decades ago, Campbell and Stanley (1966) enthusiastically argued that using a post-test only design was more robust than using a pre-test–post-test design because of the inherent difficulty in demonstrating that the sheer act of engaging in the pre-test itself did not have some impact on students' abilities when taking the post-test. Despite this perspective, a pre-test–post-test design is the dominant approach, due in large part to scientists' tacit desire to know participants' initial conditions. As a further notion worthy of comment, the two-group study obviously isn't truly an experimental design unless study participants are randomly assigned to one of several different treatment groups; when participants are not randomly assigned, this approach is more traditionally termed *quasi-experimental*. For most applications we see in the literature, random assignment of participants to treatment groups is rather unusual.

Measuring Student Attitudes, Values, and Interests

EXIT

See: *Assessing Affective Characteristics in the Schools* (2nd ed.), Lorin W. Anderson and Sid F. Bourke, Lawrence Erlbaum, 2000.

In addition to pursuing how much content students have learned, frequently called either content knowledge or conceptual understanding, AERers often want to know if students' feelings, known as affect, have been impacted by an educational intervention or experience. A detailed description of how psychologists define and study affect is a complicated and intertwined rabbit's warren, so we won't go into substantial detail here. For our present purposes, let's oversimplify it and consider that affect is a rich combination of *attitude* (how much one "likes" something), *interest* (how often someone will choose something over something else), and *value* (how important something is relative to other things).

   Conventional Likert-scale. Although hotly debated in many circles, the most common and efficient strategy to get some quantitative feel about participants' affective characteristics is to use a Likert-scale. (Likert is actually pronounced "LICK-urt," but is rarely ever pronounced correctly, including by the authors of this book, so don't worry too much about getting it right. We also often knowingly mispronounce *Tycho Brahe*, *Vega*, and *Io* and we don't get in a tizzy about that either, so you'll have to judge for yourself whether or not to commit such an academic peccadillo.) As it is most widely used, a Likert-item provides respondents with an idea and asks them to judge the degree to which they agree or disagree with the statement. Its most common form is a 5-point scale, but other scales and ideas about how best to go about this business do exist.

Hypothetical Likert-Scale Items

Please rate the degree to which you agree or disagree with the following:	Strongly Disagree	Disagree, but not strongly	Neither agree nor disagree	Agree, but not strongly	Strongly Agree
1. Astronomy is the most important discipline in all of the scientific enterprise	1	2	3	4	5
2. *Conducting AER: A Primer* is the most insightful and riveting page-turner I've read in a long time.	1	2	3	4	5

            © W. H. Freeman and Company, 2011

These variations include only labeling the end points of the continuum, forcing a positive or negative choice by not providing a middle or neutral option, and using 7- or 9-point scales. Other strategies that might surprise you are related to the tendency of some respondents to simply select "5" for everything. Some Likert-scale survey designers will insert several items that are "opposites" of each other, in which to be consistent, a respondent must mark opposite ends of the scale. For instance, one item might say something like: "Astronomy is my all-time favorite class." Another item might say: "Astronomy is my least favorite class, ever." It is relatively easy to select out and discard the surveys of respondents who answer "Strongly Agree" for these diametrically opposed items. In other cases, survey creators may attempt to ward off participant fatigue by reversing some items. Or, since many items are written in search of socially desirable results, it may be useful to insert items that are written in order to detect a negative result. For instance, if you are designing an end-of-course survey, you may want to balance a "positive" item (*e.g.*, "The course web site provided useful resources") with a "negative" item (*e.g.*, "Purchasing my *iClicker* was a poor use of my money").

If you use either of these strategies, and you are planning to average the results from many items, remember that you need to flip the scores such that all responses that you view as being desirable are given the same numerical value. Let's consider an example. Imagine one item might be something akin to, "Astronomy is my favorite subject" and then an item appearing later on the survey might read, "Astronomy is my least favorite subject." This second item is called a "reversed" question, in that the question is negatively oriented relative to the other questions. When preparing data for analysis, the fastest strategy is to take every response in this column and subtract it from the number 6 (assuming you've used a 5-point scale). In this way, an item that initially appears as a 4 in the raw data spreadsheet now magically becomes a 2 because $6 - 4 = 2$— *voila*!

Likert-scale surveys are fast and painless, and depending on which of us you are talking to, might be worth about as much effort as it takes to administer it. Recent research suggests that there can be important, and less than obvious, influences of gender, culture, and intent in how people respond, which we will discuss in a later section. Nonetheless, it is a staple in the AER toolkit and you should be aware of it. We have provided an example Likert-style attitude survey for you in Appendix B at the end of this book.

  Calculating and Reporting Scores on a Likert-scale. There are several ways to make sense of the data that result from administering a Likert-type scale. The most common strategy for analyzing and reporting Likert-scale surveys is to calculate the average score and standard deviation for each of the items, and/or the survey overall, in a manner similar to comparing the pre- and post- test scores from a content-oriented assessment. When multiple groups are being compared, the *t* test of means, as described earlier, is used to initiate the analysis, although it is certainly not the only statistical approach available. This can be done both in terms of comparing pretest to posttest, as well as comparing treatment groups to each other. When using these techniques, it is common to compare the pretest scores of each treatment group in order to demonstrate that both groups were equivalent before any educational interventions were applied.

Likert-scales, Part Two

Having said all of that, we feel compelled to say a word or three about the blithe and occasionally bizarre ways in which Likert-scales are misused in research (including some practices that we have already described in this section). First, all too often, the distributions of responses are not provided, and cannot be discerned without, and sometimes even with, the standard deviation. For example, studying a population of 20 participants, an average of 3.5 can be produced by 20 responses of "3" and 20 responses of "4." The same result can be produced by 25 participants providing a response of "5," and 5 participants providing responses of "1."

Completely Fabricated Likert-Item Score Report

Likert-scale Items where *1 = strongly disagree* and *5 = strongly agree*	Number in Group A/B	Mean Group A (Standard Deviation)	Mean Group B (Standard Deviation)	t test p-value	Statistical significance at α=.05
1. Astronomy is the most important discipline in all of the scientific enterprise	22/19	2.75 (0.25)	4.25 (0.75)	0.02	YES
2. *Introduction to AER: A Primer* is the most insightful and riveting page-turner I've read in a long time.	22/19	4.50 (0.25)	4.25 (0.75)	0.36	NO
Overall =	*N* = 41	3.63 (0.25)	4.25 (0.75)	0.06	NO

Not only are these very different scenarios, but neither represents a normalized distribution, making the use of statistics such as the average and the *t* test questionable. Experience has taught us that we really do want to know what the distribution is, as the mean of the data is nearly meaningless. One of our greatest pet peeves is to see such data reported without so much as the standard deviation. On a 5-point scale, a mean of 3 with a standard deviation of 0.5 is entirely different than a mean of 3 with a standard deviation of 1.5, so please report your standard deviation, and even better, provide us with a histogram illustrating the distribution of responses.

Further, we admit that it is perfectly possible to calculate averages and *t* test values for Likert-scale data. However, we feel compelled to point out that doing so may not make sense. The values on a Likert-scale are most often ordinal, rather than interval, in nature. This means that, although the numbers run in order from 1 to 5, for example, one cannot say what the difference is between each number. We cannot say, for instance, that the difference between a 1 and a 2 on a Likert-type scale is the same as the difference between a 2 and a 3. The numbers are "stand-ins" for the idea of difference, without being able to actually say what that difference is. We cannot assume that the differences between adjacent levels are equal. As Jamieson (2004) points out: "The average of 'fair' and 'good' is not 'fair-and-a-half'; which is true even when one assigns integers to represent 'fair' and 'good'!" For this reason, it really doesn't make any sense to calculate our means and t scores. This is an example of using a statistical test because the button exists in the program, without understanding the meaning of such an action. We see this improper use of statistics in far too many studies. We even admit to doing it ourselves, but we're not proud of it.

Looking for Changes in the Distribution of Likert-type Data

Another way to look at Likert-type data that acknowledges the ordinal nature, and the frequently non-normal distribution of the data, is to visually inspect, and statistically analyze, the distribution of frequency of the data. In order to visually inspect the distribution of students' responses to a Likert-type survey, all one needs to do is take a look at a histogram of the data. For instance, let's say that you wanted to know how a beginning-of-year instructional intervention might function to alter students' perceptions of science. Let's further say that you are working with a population of students who are unlikely to truly appreciate any of their coursework, so you decide NOT to ask them whether or not they like science class, but how they well they like science as compared to their other courses. You collect this data the first day of class, and again three weeks into an active engagement introductory science course. Upon calculating the means of the surveys, you see some change, but not as much as you might hope to see.

You pick up a copy of your very favorite AER primer, and read that it might be more fruitful to look at the frequency of responses in each category, and to run a Chi Square test to see if there is a statistical difference in the distribution of responses. You create histograms of the data, pre- and post-, and see something that is very encouraging:

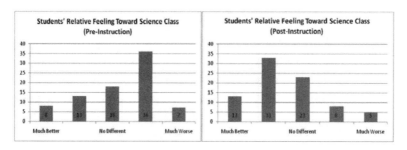

Wow! That's a pretty exciting shift, especially so early into a semester. You put the data into your favorite software program and wait with bated breath. To your gratification, a Chi Square analysis indicates that these are different distributions, with a probability of this being a chance result of less than 0.01%. According to your standards, this is a statistically significant result. Very cool. Not only are you doing something in your class that is making a difference in your students' conception of what it means to be in a science class, you've shown that you're making a difference using methodologies that make sense.

Interpreting Likert-type data in this manner takes no more time than calculating a *t* test, and it is a more appropriate way of handling ordinal data. At least one of us tries to point this out to researchers (at every available opportunity), and receives blank stares in response. It's a bit like tilting at windmills, but we include this in the hopes of positively affecting at least one study.

Correlating Attitudes and Conceptual Diagnostic Scores. As scientists, we love to wax poetically about cause-and-effect relationships. To restate what we alluded to earlier, statistics do not have the power to establish cause-and-effect relationships; cause-and-effect relationships must be argued for through non-statistical means. However, statistics can provide a quantitative feel for how variables co-vary, or *correlate*, with one another. For example, an enthusiastic AERer might wish to know how student attitudes correlate with student achievement. An easy-to-use correlation statistic found in packages such as *Excel* or *SPSS* is the *Pearson Product Moment Correlation Coefficient*, or more simply, *r*.

The Pearson correlation can be used when dealing with ratio-type data, where the ratio between each number is the same. For example, the difference between 30 and 32 on a test is the same as the difference between scoring 90 and 92 on a test. They each have a difference of two, and two is the same no matter where you are on the scale. If you are using non-ratio data, like the ordinal data that one might collect from a Likert-type scale, the Spearman correlation is the correct statistical test to use. (More on Likert-type scales later.)

The Pearson correlation value ranges from -1.0 to $+1.0$, where zero implies no relationship between the two variables being compared. As the values move away from zero, the relationship between the variables can be considered to be stronger, with positive values indicating a direct relationship and negative values an indirect relationship. AERers most often report the r^2-value, which is a measure of the proportion of the variance explained by the relationship. So, as a completely hypothetical example, if correlation between student attitudes and conceptual diagnostic scores happened to be $r = 0.8$, then $r^2 = 0.64$. 64% of the variance in student attitudes can be explained by conceptual diagnostic scores. In a gross overgeneralization, r^2 values in the range of 0.50 to 0.80 would be considered quite high in many educational applications.

Caveat Emptor

Without question, the emerging AERer is immediately attracted to quantitative and statistical methods of research. Most individuals who are trained in astronomy already have some facility with describing the world numerically. However, one can apply statistics to almost anything and simply having numbers that match one's expectation can set a researcher up to make assertions and interpretations that might not hold up under closer scrutiny. An abbreviated list of the many things that can jeopardize the validity of one's study includes:

Placebo Effect – This occurs when participants believe or perceive they have been given a better educational experience than they would normally. This is sometimes known as the *participant-expectancy effect*, and it results in students engaging in the course non-standard ways, and potentially increasing their scores due to reasons other than the instructional intervention.

Nocebo Effect – This is the opposite of the *placebo effect*, and occurs when participants think they have been disadvantaged by a particular educational intervention, or by not receiving the "special treatment" given to another group. In such cases, students may perform more poorly than they would otherwise. On the other hand, such a perception can also cause the John Henry Effect, in which students who believe that they are in the underdog group go out of their way to show that they can overcome their disadvantage. Either way there are variables at work that are likely unrelated to the intervention being studied.

Hawthorne Effect – This is a well-recognized effect in which people actually demonstrate short-lived, but measurable, improvements, simply because they know they are being observed.

Pygmalion Effect – This is also known as the *teacher-expectancy effect* or *observer-expectancy effect*. It can occur when you, the researcher, think you see improvement simply because you expected to. This is certainly related to our own desires to believe that we are helping our students, our need to believe that we have not wasted our time in conducting a study, and job and ego pressures that we might feel to publish our findings. The result can often be that the researcher is so committed to seeing a change that they unknowingly (or sometimes knowingly) provide additional interventions and introduce unaccounted for variables into the study.

All of these biases are related to what we like to call the **Feynman Effect**. This is in reference to a well-known quip from Richard Feynman: "The first principle is that you must not fool yourself, and you are the easiest person to fool." We've never heard the term "Feynman Effect" used by anyone else in this way, but we like it and hope that it catches on. It is so easy to fool ourselves when we are engaged in research, particularly educational research, and very particularly educational research in our own classrooms. We have to be wary.

We are apt to presume that statistical methods are objective and therefore the most valuable approach in conducting AER. However, upon closer inspection, this perspective isn't as solid as one might initially think. In the end, one finds that, as powerful and convincing as quantitative statistical methods seem to be, the choices one makes, such as which instruments to apply, when, to whom, and who exactly should administer them, can be highly capricious and subjective. These choices are influenced by our philosophical approach, expertise, prior knowledge, and to be honest, ease of execution. In the next chapter, we present some important AER strategies and tools that can ground-truth your research and have the potential to provide even greater insight into the teaching and learning of astronomy.

CHAPTER 5
Qualitative and Interpretive Methods: *Revealing Understanding*

In astronomy education research (AER), just as in traditional astronomy research, different questions require different approaches. Even though spectroscopy is a commonly used tool in astronomy research, it doesn't yield fruitful information in every scenario. For example, an astronomer interested in determining the diameter of a Kuiper Belt Object isn't going to find much useful in observing with an Echelle Spectrograph, no matter how much she wishes it were otherwise. In much the same way, if an astronomy professor who wants to understand which learning experiences help move her students along a pathway from being a student to becoming an astronomer, a test of students' memorization of constellation names isn't going to yield much useful information. In fact, to get at *how* students' mature from a student to a scientist so that you can help them along their path, one might find that no written test or survey is going to provide insight. In such a case, we have to turn to different and more appropriate instrumentation—qualitative, or interpretive, methods.

In contrast to the quantitative data collection and analysis strategies presented in Chapter 4, qualitative studies tend to incorporate methodologies that include the collection of extensive, detailed data about a small number of subjects (i.e., participants, events, scenarios). These qualitative approaches are intentionally designed not simply to describe, but actually to *interpret*. For example, one might describe a room as having dimensions of 15' long × 20' wide × 10' tall with low wattage, incandescent lighting, light brown walls and dark brown carpeted floors. Alternatively, a description might propose that the same room was intimate for a small gathering, absent furniture to encourage movement and mingling by its occupants, and dimly lit to promote a relaxing mood with diverse conversation. Depending on the purpose for making the description in the first place, one approach is fundamentally different, but not necessarily better, than the other.

As you might imagine, such data is often verbal, whether oral or written, or observational in nature and is not always easily represented with numbers or analyzed through traditional statistical measures. This does not mean that a study is qualitative in nature, if the data used are non-numerical. Interviews and sketches can be coded, counted, and statistically analyzed in much the same way as multiple-choice test data. From a theoretical point of view, there is little difference between using a multiple-choice test or an open-ended survey in which the coding of the responses is pre-determined. For this reason, we've begun using the term *interpretive* to describe work that has its "central research interest in human meaning in social life," particularly meaning as determined by the subjects (Erickson, 1986, p. 119).

One might think to themselves, "Why, that seems awfully touchy-feely, and not at all empirical!" By this, we think that people mean that they believe this kind of work is not "scientific" and lacks the systematic protocols that characterize empirical work. This is a sentiment that qualitative/interpretive researchers often hear from those who come from a more positivist orientation. At moments like these, we try to remind our readers that even within the "hard sciences," the conventions of what can be considered evidence, and what constitutes appropriate research methodology, varies widely. In the end, what really matters is that one uses the correct methodology for the research study. Over the past few centuries, conventions related to studying people have been constructed, and interpretive research falls within accepted (and encouraged) methodologies.

With regard to the shoddiness of some qualitative/interpretive work...well, we find that we cannot argue. Of course, we also think the same of some statistically oriented, quantitative work that we encounter. Shoddy work is everywhere, but really well done interpretive work can tell us things about people that we could never ferret out using other methods. This is primarily because interpretive work focuses its attentions on the complexities of human interactions, and on the meaning that the people involved make from their circumstances. These are the kinds of things that do not lend themselves to "checking the box," "filling in the survey," or indicating where your opinion falls on the "Strongly Agree" to "Strongly Disagree" spectrum. Being human is far too messy to be understood using only those methods. Interpretive research is intended to make meaning out of that messiness.

Interpretive Research Addresses Particular Types of Questions	
Interpretive Domain	*Examples*
• What is the social interaction in the situation?	• What are the unwritten "rules of engagement" in a think-pair-share experience?
• What is the specific structure of the occurrence?	• What's the nature of the relationship between undergraduate astronomy majors and their research advisors?
• What meaning do the people involved make of the situation?	• Some students adhere to an astrology belief system. What impacts to student learning occur when their ASTRO 101 professor makes light of their belief system?
• It seems that the existing model doesn't explain the situation— what other variables might be at work?	• We've been told that women and minorities don't favor "science and engineering," yet our afterschool robotics club is filled with girls and minority kids. How can we explain this divergence from the common wisdom?
• How does what is happening here relate to what is going on in the larger setting?	• How does using lab materials that are based on social-constructivism function to impact students' attitudes towards the larger scientific enterprise?
• The situation has superficial similarities to another situation—in what ways are they the same or different?	• I'm using clickers in my lecture and having much better results than my research partner across town. I don't think that my implementation is any better, so what's going on here?

General Characteristics of Qualitative Research

Interpretive research tends to occur in the natural settings of the participants. It is most often characterized by extensive field notes that are subsequently coded and analyzed in various ways. For example, if you want to understand the characteristics of conversations that take place between students using *think-pair-share* in an introductory astronomy course, you would need to systematically observe them in that class while the discussion is taking place. And, more importantly, the qualitative researcher would not just note precisely what students were saying, but try to uncover and describe why they were saying what they were saying and how they were saying it. At other times, qualitative research may be interactive, in that the researcher is communicating with the participants in an ongoing manner, for example, within the classroom or through an interview, rather than observing them from a distance.

Unlike quantitative research, qualitative studies do not usually have a hypothesis to be tested or a prediction to confirm. In this we think that astronomers might have an easier time of adjusting to interpretive research than researchers in other fields. For instance, the scientific discipline that we are most often associated with is astronomy, an observational, rather than an experimental science. As one colleague has said, "I've been an astronomer for nearly thirty years, and I've never written a hypothesis. I've never run a controlled experiment." Instead, an astronomer, much like an interpretive researcher, combs through and cleans data, many times in an iterative fashion, continually looking for new evidence and continually striving to better understand the data they have collected.

Data That Can Be Used Both Qualitatively and Quantitatively

Interpretive methods might initially seem pretty foreign to an astronomer, and it is unlikely that you are interested in using the extreme end of the quantitative to qualitative continuum in your first project. So we thought we'd start with some examples of data that can be approached from both qualitative and, if appropriate to answer your research question, quantitative analysis perspectives.

Let's imagine that you want to know something about your students' ideas about stellar evolution. One way you might do this is through an open-ended or "essay" question; perhaps such a question is already given on your unit test. Using some kind of rubric or guideline to assign a score to each response, such as 8 out of 10 points, is a reasonable process for grading your test, and the mean score could be used to get a sense of the extent to which the class as a whole is meeting expectations in this topic area.

But this process doesn't tell you much about what aspects of stellar evolution your students seem to understand well, and where and why they are still having difficulty. A different analysis technique might be to read through the students' responses, documenting recurring themes as you read. Doing this in an iterative process until you can identify all major themes *and* identify evidence of any themes present in a given response is a common way of getting a handle on the ideas that the students have (or don't have). The themes that emerge from the data will be particular to that data set, rather than being matched to preconceived notions, categories, or theories. This process is commonly called *grounded theory,* in which the results emerge from the data.

In the case of essay questions another approach is to iteratively code for recurring themes (qualitative analysis) but then to assess the frequency of those themes (quantitative analysis). It is also possible that this type of analysis is completed with other qualitative data. For instance, after completing an initial analysis of the data, you might choose to interview a selection of students. You might choose to select the students randomly, as dictated by a more positivist approach. Or, you might choose to purposefully select interview subjects based on criteria: their deviation from the norm, the richness of their initial response, or some other value. Whatever criteria you use, in interpretive research the point is to collect data that will provide dense evidence that addresses your research question.

The important issue is *what are you trying to find out, and how will these data and analysis strategies help you answer the question?* Without alignment of your questions, data, and analysis strategies, the research may lack meaning. Most researchers bring all of this together by carefully thinking about what frames their thinking—in other words, by identifying their theoretical framework.

Theoretical Frameworks

A critical component of qualitative research is the *theoretical framework* you use. This is the lens through which you approach your research, and helps the consumer understand what kind of assumptions and methodologies you will use. Identifying your theoretical framework(s) is frequently required when proposing or publishing; this is becoming a critical issue even for quantitative studies, which traditionally have been about hypothesis testing and previously were often considered atheoretical.

> **EXIT** ▶
>
> See: *Theoretical Frameworks for Research in Chemistry/Science Education*, George M. Bodner & MaryKay Orgill, Pearson/Prentice Hall 2007

Selecting your theoretical framework is much like selecting your wavelength regime and instrument for astronomical observations. No matter what wavelength or instrument you use, you will most likely "see" something, but the lens you use will greatly influence what you are able to see. You can observe the Cassiopeia A remnant using Chandra in X-ray, MDM in optical, or the National Radio Astronomy Observatory, and you will see something interesting each time, but you won't see the same thing. This doesn't make astronomy a wooly discipline, or unreliable. It just means that, even in a cold, hard field like astronomy, the way you choose to examine the world around you has profound influences on your results.

So what would this look like in AER? As just one example, a study that investigates female astronomy majors' experiences in their senior capstone project might use a theoretical framework of *feminism,* which assumes that some aspects of scientists are biased against women and that these aspects should be identified and explained in order to alleviate the problem. Feminism assumes that the researcher in this case cannot be a distant, objective observer and that your own perspective (whether it is as a White male or a Latina female, for example) will have an impact upon the way in which you view the data. Alternatively, one might address the same issue using an *identity* framework, in which one believes that individuals construct a sense of self by constantly comparing their interactions with the world to their conception of the "possible self," pruning interactions until the two data sets are commiserate. Just like using different instruments to view a supernova remnant, using these two frameworks to bear on a single data source will provide different results.

Conventional Qualitative and Interpretive Studies

Ok, let's say you're interested in finding out even more about a participant's learning than can be understood by using one of the methods above—perhaps you want to know not only *what* a student knows but *how* she makes a mental model to explain a given astronomical situation. What kind of data would you need to collect?

The primary data collection strategy in this case is to *interview* the student(s). Spending 30-60 minutes in a semi-structured interview (i.e., you have some questions determined in advance, but you also are willing to ask additional, unanticipated questions as they become needed as a result of the participant's responses) could yield great insights into the thinking of the student as she reasons through, say, the life cycles of stars of various initial mass. During the interview, you may ask the student (or she may volunteer) to draw a diagram or picture of her thoughts; this *artifact* then becomes an additional source of data.

In the present sense, there are essentially four core sources of qualitative data, of which interviews are one. The others—documents, observations, and audiovisual materials—can each offer yet another source of rich detail. The analysis of interview transcripts, documents, observational field notes, and the like would include reviewing all of the pieces for commonalities, which will allow you to develop themes and identify individual pieces of evidence (e.g., quotations from the interview) to support them. A selection of other forms of data are listed in the box that follows.

Example Data Collection Strategies for Interpretive Research

- *Field notes from observing as an outsider*
- *Field notes from observing as a participant*
- *Notes from unstructured, open-ended interviews*
- *Coded transcripts from recorded unstructured, open-ended interviews*
- *Notes or transcripts from semi-structured or structured interviews*
- *Notes or transcripts from focus group interviews*
- *Summaries from written journals or log books*
- *Email correspondence collected from or among participants*
- *Photographs or videos collected from participants*
- *Photographs or videos made by the researcher*
- *Public documents, blogs, websites, meeting minutes, or newspapers*
- *Discussion diagrams charting who said what to whom and how often*
- *Trace evidence, such as footprints or cell phone records*
- *Artifacts, such as class notes, exams, PowerPoint slides, or websites*

Going to the "Fringes" in Interpretive Research

See: *Qualitative Inquiry and Research Design: Choosing Among Five Traditions*, John W. Creswell, Sage Publications, 1998.

Creswell, referenced in the nearby EXIT box, identifies "five traditions" of interpretive study, all of which are relevant to understanding the nature of teaching and learning in astronomy. Each of these methods is far beyond the usual tools of an astronomer, and represents work that is beyond what the casual education researcher should take on. However, as you find that the methods you have been using are not giving you the kind of insight you need to understand the phenomenon that you are studying, you might want to join forces with an experienced education researcher to give one of these a try. Remember, which approach you use depends on what you want to learn, because each provides slightly different, and potentially more or less useful, information. In general, when you are reporting this type of research, you need to describe which of the traditions you are using, so we briefly describe them here.

Biography. Most of us have read a biography sometime in our schooling years, but perhaps you've never thought about creating one yourself in support of AER. From an analytical perspective, a researcher presenting a biography tells a chronological story of or reports on patterns from conversations with a single, distinctive individual. More than simply recording and reporting a single individual's life or learning experiences, a researcher conducting a biography must situate the events in a larger context and, most importantly, describes lessons learned in conducting the study. The most common data sources for a biography are interviews or collected correspondence. A biographical approach might work well for understanding how a well-respected astronomy professor's philosophy and approach to teaching has evolved over time due to a specific set of events.

Ethnography. An ethnographic study answers research questions about a particular group or culture. As such, a common characteristic of an ethnographic study is "participant observation" as part of field research. In other words, ethnography is a study of a culture by an outsider, who has to try to find ways to become part of the culture in order to study it. The most common data sources are observational notes or in-depth interviews with members of the targeted culture. An AER example would be a researcher living and working in a school in the Navajo Nation, trying to understand the Navajo conception of astronomy. Another example would be an attempt to understand the functioning of an informal study group of African-American women that extemporaneously forms at a coffee shop in supportive preparation for an ASTRO 101 final exam

Phenomenology. A phenomenology is a study of a group of individuals who have a shared experience. Accordingly, one might describe a phenomenological study as one that provides a narrative description of how a group of people found meaning in a particular event or phenomena. Because it is steeped in "meaning," a phenomenology intentionally emphasizes uncovering people's subjective experiences and interpretations of the world. In other words, a phenomenological study endeavors to understand and report how others see the world. For an AERer, a phenomenology could pursue how female students see their roles and responsibilities, and that of their instructor, in the ASTRO 101 classroom when required to work through highly scripted tutorials during class in collaborative groups.

Case Study. A case study is an in-depth description of a particular individual or a specific context, often in the form of a vignette. Probably the most famous case studies were those done on a few individuals by Sigmund Freud in developing his theories of psychoanalysis and those done by Jean Piaget of his children in developing his theories of how children develop and mature. An important part of research done in a case study is to identify if the case is typical, atypical, or extreme, compared to some described norm. As an example, an AER case study might pursue the case of how a particular first-year assistant professor balances teaching and research and what support mechanisms he or she believes to be most helpful.

Grounded Theory

See: *Basics of Qualitative Research: Techniques for Developing Grounded Theory* (3rd ed.), Juliet M. Corbin and Anselm Strauss, Sage Publications, 2007.

A grounded theory approach, described briefly earlier, uses the theoretical perspective that what is going on is unknown *a priori*. As such, no sort of hypothesis testing methodology makes sense. Rather, a researcher using a grounded theory approach would let themes emerge from the data. Then, once themes are identified, a researcher using a grounded theory approach would intentionally go back into the data set and look for the existence of disconfirming cases to establish the results of the study. The four basic steps of this process are briefly described in the box below. Grounded theory is an incredibly powerful approach and has become commonplace throughout the interpretative research realm.

Data Analysis Approach in Grounded Theory

1. *Open Coding – forming of initial categories, where each category has properties along a continuum of possibilities*
2. *Axial Coding – reassembling the data in new categories looking for causal conditions*
3. *Selective Coding – preparing hypotheses or propositions and a "story line" based on axial coding*
4. *Testing the Propositions – Purposeful collecting of confirming and disconfirming cases from the data, or a new data set, that test the propositions from selective coding*

Talking with Learners: *How Does One Go About Scheduling, Conducting, and Analyzing Interviews and Focus Groups?*

A common characteristic of astronomy education researchers is that they seem to be individuals who like people. And, talking directly with professors and students yields a tremendous amount of highly valuable data. But, it's not just walking up to someone and saying, "Hey, how did you learn that?"; rather, there are nuanced ways of interacting with people that yield valid and reliable data—data that wasn't unintentionally tainted by telegraphing the answers you were hoping for rather than what is really going on.

Finding Participants and Scheduling an Interview. One of the things that you'll need to decide as part of the research design is *who* you will interview. If you're interested in the experiences of female science majors, that will automatically set up some boundaries for your potential participants. If you want to investigate a phenomenon from the perspective of a variety of students, you might try a purposeful strategy that aims to recruit an equal number of students who earned high, middle, and low scores on their first exam. As always, how you go about recruiting students will depend on your research questions. In reality, you should also recognize that there can be a *self-selection effect,* in that it is sometimes the more motivated (and possibly higher achieving) students who tend to volunteer, or that compensation might alter the volunteer pool; noting this limitation, if it applies, is always a good idea when you present your results.

If you have a Ph.D., you might recall that one of the most challenging tasks was arranging for all of your committee members to be in the same room at the same time. We hate to be the ones to tell you this, but you need to be prepared: Finding and convincing interviewees to actually show up is an arduous challenge. Compensation for the participants' time—in the form of extra credit points, donuts, or coffee shop gift cards, as just a couple of examples—can help, but must be approved by your IRB.

Designing an Interview Script. In deciding what you want to ask your interviewee, you have two critically important and simultaneously competing issues to deal with. The first issue is to clearly define precisely what it is you want to learn by talking with your interviewee. This means before you interview someone, you should write down what evidence you are looking for and what some counter evidence might look like. Then, carefully and intentionally phrase your questions to target finding evidence. This is far easier said than done and you should plan on trying out a few test interview scripts before you have exhausted all of your potential interviewees. Over time you will find out how to better improve your scripts.

Sample Individual Interview Script (Abbreviated)

BACKGROUND ABOUT STUDENT
- *Please tell me about your background in science, such as describing any science classes you have taken or any informal science interests that you have pursued.*
- *Have you been taught anything about stars? Please describe the lesson/context/situation as best you can remember.*

STAR PROPERTIES
- *Tell me what you think a star is.*
- *How did it (the star) end up that way?*
- *Could you draw a picture of what it (the area) looked like before it was a star?*
- *Show me another picture of an in-between time (between your last picture and when it is a star).*

MAKING COMPARISONS BETWEEN OBJECTS
- *Is a star different from a planet? How so/why not?*
- *Is the Sun the same thing as a star? Why/why not?*

At the same time, our experience is that interviewees generally want to be helpful. While this initially sounds like a useful thing, it can also be a sticky problem. What happens in an interview is that interviewees will actively try to figure out what you *hope* that they will say. Strange as this might sound, interviewees seem to do this subconsciously by looking at your body language, facial expressions, and subtleties in the way you word your questions—they are trying to read between the lines of what hidden agenda you might have. Now, we could wax poetically all day about why this might be so, but it wouldn't be very productive. So, let's just leave it as it is; you have to be abundantly cautious about telegraphing your underlying research agenda or hypothesis testing to your interviewees lest you risk ending up with highly biased data.

Focus Groups Provide Different Insights. A powerful variation on the individual one-on-one interview is to bring small groups of 4–10 people together and do a group interview. Pioneered in the commercial world of product marketing research (and now used frequently to shape political campaign messages to be palatable), *focus groups* take advantage of the social interactions to learn even more about a particular issue. The beauty of focus groups is that, when run well, participants don't simply talk to the interviewer, but rather talk with other participants. The advantage of this is that, during normal social discourse, people are apt to say things like, "Oh, I've never thought of that before" or, more insightfully, "Now that you mention it, that is precisely what I was thinking too." Focus groups rely on human beings' inclination to hop on and "piggy back" on one another's ideas. This is tremendously advantageous in that the interviewees themselves help identify and highlight important and cross-cutting themes in the discussion in ways that might not naturally occur in a one-on-one interview.

Analyzing Notes and Transcripts. As there are many types of interpretive research, there are many ways to analyze the data. Books and books have been written on the topic, indicating that there is no way that we can be thorough in this small book. Having said that, the analysis techniques used in grounded theory research are fairly flexible, and can apply to many kinds of questions. Therefore, we share one approach with you, knowing that we should be bracing for the onslaught of researchers who will want to correct what we write here (by which we mean, they would like us to provide instructions for *their* favorite analysis technique).

Coding the Data. In interpretive research, and particularly grounded theory, it is common practice to begin analysis by scanning the data sources for obvious connections between the research questions and the data. Some might insist that this step unduly influences the researcher's perceptions of the data, and others dismiss this step as common sense. We like to think that there is no way that one cannot approach a data set on a topic that one has put a great deal of thought and literature review time into without some preconceptions, and we're not sure one should try. Having some expertise in the field allows the researcher to see things in the data that an outsider might not be able to see. Awareness of your preconceptions is enough. Moreover, the acts of cleaning, transcribing, reading, and re-reading the data are considered by many to be essential to the act of interpretation.

Sample Focus Group Interview Script (Abbreviated)

Introduction:

❑ *My name is Draco Saturnalia and I am an astronomy professor here.*

❑ *Your professor has asked me to come and chat with you today to get an understanding of the <u>strengths and weaknesses</u> of how they have designed this class so that they can make improvements in the future.*

❑ *I would like to record our conversation so that I can prepare accurate notes to share anonymously with your professor and my colleagues after grades have been submitted for this semester. In no way will specific comments you make ever be associated with your name and no one other than me will listen to this recording. <u>Your participation is voluntary.</u>*

> ❏ *My intention is that you'll have a conversation with one another, not just back and forth with me, and that one person's ideas will spark thoughts in others. You don't need to raise your hand to speak. If you don't have any questions or concerns, allow me to turn on the recorder and we'll begin.*

<u>Questions:</u>

ONE: Since I don't know anything about the way this course is designed, can you describe how it is different than other similar courses you have taken? *How is it similar? How is it different? Can you tell me a little bit more about that?*

TWO: Your professor mentioned that he has been assigning activities and/or projects in this class that are different from the typical homework set. Can you describe how they are different, if at all? *What did these tasks help you learn?*

THREE: What did you like best, if anything, about these activities/projects? *What and how did these tasks help you learn?*

FOUR: What did you dislike most about these projects/activities? *How could they be changed to help you learn astronomy better?*

FIVE: Is there anything else that the course instructors need to know as they think about redesigning the class for the future? *What should they keep and what should they change?*

Coding and Memoing. As we have indicated, the basis of analysis of the grounded theory approach is to read and re-read the textual database—interview notes and quotes, yearly surveys, interviews, etc.—which allows the researcher to discover and label variables. These variables are called *categories, concepts* or *properties*. The ability to perceive variables and relationships is termed "theoretical sensitivity" (Glaser, 1978) and is affected by a number of things, including the researcher's reading of the literature, prior knowledge, life experiences, and the use of techniques designed to enhance sensitivity.

Two of the most important techniques available in grounded theory are *coding* and *memoing*. The first technique, coding, is a means by which data are organized into categories according to their properties. In the process of *open coding*, items are identified from the data and are defined or arranged according to their various properties. The magnitude or dimension of the property may be described. As open coding proceeds, the data are fractured and compared for similarities and differences. Events and ideas that seem to be related to each other are grouped as categories. During open coding, data are scrutinized through constant comparative analysis (Strauss & Corbin, 1990). This process frequently results in re-categorization, wherein some categories are broken into sub-categories, data are moved from one category to another, or data from multiple existing categories are moved into a new group. The result is a set of concepts that "depict the problems, issues and matters that are important to those being studied" (Strauss & Corbin, 1990). *Axial coding* is the process of reintegrating the categories in such a way as to begin to explain the phenomena. Axial coding looks at how categories relate to each other, and how certain properties may cross-cut categories.

These forms of coding do not occur sequentially, rather, the processes are highly interwoven, while at the same time, theory is evolving. In order to manage the large cognitive load required in grounded theory, researchers rely upon *memoing*. Memoing is simply the act of making notes to keep track of ideas that occur. This can occur in the margins of a sheet of notes, on paper note cards, or in a computer program. The method is not important. In a similar way, coding can occur in the margins, on small snippets of paper that have been cut out of larger pieces of text and color-coded, or by using software. Dale Baker described her analysis for her seminal work "Letting Girls Speak out about Science" as consisting of a multitude of scraps of colored paper being shuffled around on her living room floor until they made sense (2008). In the case of much of our work, theory generation was confined to a spiral notebook in which the coding was copied over and over again as it evolved. Other researchers are a bit fancier in the tools that they use, relying on software, such as NVIVO8, to manage the analysis process.

In the final steps of analysis, the data are integrated to develop a description of "what is going on here" (Strauss & Corbin, 1990). In grounded theory, as an example, this theoretical conception is rarely represented by one lone case, or by the words of a single participant.

Each category's description, and the description of the cross-cutting category, usually represents the point of view of many subjects. Findings of this nature are not represented as raw data, but as narrative descriptions.

A Final Note

Like all good research, equally important as *what* you do is *how* you do it—and that means having a good alignment between your research questions, data, analysis strategies, and interpretations. Both qualitative and quantitative strategies can lead to valuable contributions to the field, and one is not necessarily better than the other. Instead, it's all about which is better for what you're trying to find out.

CHAPTER 6
Publishing: *We Celebrate Galileo Because He Bothered to Write it Down*

E ducation research is no different than astronomical research in the "publish or perish" perspective. But you've probably noticed there aren't any AER articles in the *Astrophysical Journal.* So how do you go about disseminating the results of your work in AER?

Presenting at National Conferences

Ok, so there's no AER published in *ApJ,* but there *are* education posters and talks at AAS and International Astronomical Union (IAU) scientific meetings, and there's a good chance you're already planning to go to one of those. Go check out the posters; sit in on one of the education sessions and see what they're like. You'll notice a lot of similarities to the science sessions you're used to, but you may notice a few key differences as well.

As is the case with science presentations, you'll find AERers laying out their problem, data, analysis, and conclusions. There should be an emphasis on using evidence to support their claims. There will be references to earlier work, and possibly outstanding questions and directions for the future.

We've noticed two key differences between astronomy and AER presentations (whether oral or poster). First is that AER tends to have a greater amount of time (or space) spent on justifying the question or problem that is being addressed by the study. We haven't yet figured out why this is the case, but two possibilities immediately come to mind. One is that because the audience is coming from a science rather than an education background, they may not be as informed about the background situation or assumptions and so the presenter spends more time familiarizing the audience with the problem. Another is that within astronomy, and particularly within the subtopics that are the basis for the different sessions as AAS, the outstanding problems and questions are generally well-established and known by the community (and, in many cases, the idea has already been vetted by peers when the researcher applied for telescope time). AER, being relatively new, is still quite broad and so there is not the same extent of agreement about researchable questions.

The second difference we've noticed is the willingness of audience members to ask questions and/or comment upon your research. If we were to attend a session on Lyman α galaxies, we'd probably wait and ask our questions of a trusted colleague or the presenter later, because we are far from being expert in that area and might be reluctant to display our ignorance for everyone to see. However, everyone has been to school; many astronomers have also been required to teach at some point. We all seem to feel that we know what is good and bad about education (though this knowledge may be opinion and have little to do with previous research results). As a result, people seem to be more comfortable offering suggestions and criticisms, whether they are really justified or not, in an education session. They may offer as evidence an anecdotal situation that they experienced that goes against your results, but this is not the same as offering contradictory *data*.

So far we've assumed that your presentation is about a research study, in which you will discuss a project that has involved some kind of systematic data collection and analysis in order to answer a specific question. You may find, however, that education sessions contain other presentations as well, such as those that describe a program or resource. These types of projects will be discussed further in Chapter 7.

Astronomers are most familiar with AAS, but we should mention some other good conferences as well. The main goal of the ASP is to support education and public outreach related to astronomy, and research presentations are becoming increasingly popular and valued. The American Association of Physics Teachers (AAPT) also has sessions relating to astronomy education, and has a large and well-established physic education research (PER) group as well. Finally, the American

Geophysical Union (AGU) and the Geological Society of America (GSA) both have a large number of education sessions at their annual meetings. At both AAS and AGU, you are limited to being first author on a single paper *unless* one of the two is an education talk or poster.

Finally, you should also be aware that AER can be presented through education research societies, such as the National Association for Research in Science Teaching (NARST) or the American Educational Research Association (AERA). It is important to note that while the previous organizations we've discussed have open contributed presentations (i.e., only short abstracts are required and everyone's abstract is accepted), NARST and AERA meetings require much larger, more detailed abstracts which are refereed through a double-blind system. For their 2008 meeting, NARST received approximately 850 proposal submissions and accepted just over 600. AERA is a gigantic meeting, covering the entire spectrum of education research, and last year had more than 12,000 proposals submitted!

Publishing in Journals

In addition to presenting your work at a professional conference, you probably will want to publish your results in a refereed journal. The premier vehicle for publishing AER is the all-electronic *Astronomy Education Review* (http://aer.aip.org). This journal is cosponsored and coedited by the AAS and the ASP. Other journals to consider as publication venues include, but are certainly not limited to, *American Journal of Physics*, *Journal of Geoscience Education*, *Journal of College Science Teaching,* and *The Physics Teacher.*

An article in one of these journals would be pretty similar to astronomical research publications, in the sense of major components such as an introduction, background/literature review, methods, results, discussion, and conclusions. AER articles may have fewer authors (at least there aren't yet huge collaborations, such as those associated with WMAP as just one example) and be longer in length. In particular, the introduction and literature review sections may be bigger than those in an astronomy publication. As was discussed in the national conferences section earlier, the need to justify the research project is larger. Each journal may have its own citation format (and when you start to get to more traditional education journals, this will usually be APA style). At this point there are no mandatory page charges for any of the five journals listed earlier, though *Journal of Geoscience Education* has optional page charges. Given the state of the publishing economy, it is not clear how long this phenomenon will last with these or other education journals.

Some Great Journals Astronomers Might Be Caught Reading	Some Great Journals Astronomers Probably Would NOT Be Caught Reading... But Should
• *Astronomy Education Review*	• *Journal of Research in Science Teaching*
• *The Physics Teacher*	• *International Journal of Science Education*
• *American Journal of Physics*	• *School Science and Mathematics*
• *Journal of College Science Teaching*	• *Journal of Learning Sciences*
• *Journal of Geoscience Education*	• *American Educational Research Journal*
• *Physics Education*	• *Science Education*

U p to this point, we have focused on helping you conduct and disseminate <u>research</u> on astronomy education. As in traditional astronomy, research for our purposes means adhering to the experimental and observational characteristics of science as best as possible. There are, however, some important and time-consuming endeavors that fall outside our definition of research. To be clear, these activities have a vital role in the overall astronomy education and outreach enterprise and are worthy of one's time and energy—they just don't meet our definition of research even though many do use similar methodologies and sometimes even the same instruments and publication venues. As such, we believe it is important to share with you what they are and how you might approach these scholarly activities.

Program Descriptions and Evaluations: What We Did with Our EPO Budget Allocation?

At the present time, the vast majority of contributed oral and poster presentations at scientific conferences are not actually AER *per se*, but rather *program descriptions*. Many funding entities require award recipients to share descriptions of what programs they have run, which products they have created, lessons learned, and, seemingly most importantly, who and how many individuals were impacted. To receive government funding, some education and outreach project directors need to meticulously track their participants' U.S. postal zip codes so they can disaggregate their impact and report their specific geographic impacts to legislators, who are responsible for providing services to their constituents. In some cases, projects use a creative process informally known as "project math" to generously estimate their impacts. For example, consider a hypothetical project that conducted a teaching workshop for 20 high school teachers over the summer. The project description might propose it has a far-reaching impact because each of those 20 high school teachers teach 150 students every year and, over a 30-year teaching career, this project would impact 20 teachers × 150 students × 30 years with a resulting impact of nearly 100,000 students! A fantastic claim indeed!

An important part of a project description is the overall, final (summative) evaluation of the project's worth. Such an evaluation would determine to what extent a project reached its goals and objectives, and what impact the project had

> **EXIT**
>
> See: "Finding the forest amid the trees: Tools for evaluating astronomy education and public outreach projects," *Astronomy Education Review*, **3**(2), Janelle M. Bailey and Timothy F. Slater, 2005.

on its participants. A pointer to one AER-related resource on how to plan for and carry out an evaluation is listed in the nearby EXIT sign.

To reiterate what we hope you believe we know about this, summative evaluations do not fit into our typical definition of research because they typically do not use a carefully structured methodology to answer a research question situated in the larger educational domain. This doesn't mean they aren't important. To the contrary, these projects sometimes provide launching points for important research questions worth pursing using AER approaches. As such, project description reports are something you should peruse with an eye for research opportunities when attending a professional society conference.

How a Rather Typical Program Evaluation of an Astronomy REU Turned Into an Insightful AER Study

For seven consecutive years, undergraduates attending an NSF-sponsored research experience for an undergraduate (REU) program in astronomy were interviewed at the end of their summer experience as part of a project evaluation effort. All of the alumni were contacted every year thereafter to obtain a "where they are now" description. Sometime later, a graduate student considered all of the annual program descriptions with an eye for, "Did the students' positive or negative report of their REU experience correlate with their persistence to stay in science career pathways?" This data mining, or secondary analysis, process resulted in a dissertation study of how female astronomy majors eventually decide to stay in or leave science careers. The study demonstrated that only scant evidence exists that REU programs have any influence on students' intention of pursuing science careers. Given that this program is intended to foster retention and education in STEM, a second study related to the educational function of the REU was conducted, leading to the finding that, for top-tier women in astronomy, the REU does not appear to provide meaningful educational value.

Action Research Reports

The "gateway drug" into a career in AER may be something known as *action research*[1]. As we look at how professors learn to teach, we find that an enormous amount of what constitutes successful instruction gets passed informally from experienced faculty members to newer faculty members as folk stories. Most astronomers teaching in academia cannot

---

[1] Portions of this description and included case studies were adapted with permission from a larger article (Adams & Slater, 1998).

afford the time or expense to find and use highly validated cognitive instruments, randomly assigned student control groups, psychometric item response analysis, or other hallmarks of contemporary education research to determine exactly what individual instructional activities are working or not working in their own classrooms. Further, when astronomers discuss teaching with colleagues, it is in the common language of their profession and not in the specialized language of the educational researcher. For example, *factor analysis* from the world of education research makes a lot more sense to scientists when described as an *eigenvalue problem* from the domain of astronomy, even though the underlying mathematical principles are essentially identical.

> **EXIT** ▷
>
> See: *Action Research: A Guide for the Teacher Researcher* (2nd ed.), Geoff E. Mills, Prentice Hall Publishing, 2003.

Experienced AERers tend to develop bilingual skills, using one set of vocabulary when talking to astronomers and another when talking to science educators, even when it is the same project.

At its core, action research provides a framework for faculty-led inquiry and dissemination aimed specifically at enhancing the learning environment. In general, there are six key questions that provide the structure of action research methods and results. These questions can be abbreviated as: what did the pupils actually do; what were they learning; how worthwhile was it; what did the teacher/researcher do; what did the teacher/researcher learn; and what will the teacher/researcher do now to maximize the teaching and learning benefits and minimize the disadvantages or inconveniences?

The two primary characteristics of action research that distinguish it from AER are that it is conducted by active participants in the teaching/learning process and that it is expressed in the language of its practitioners. When conducting action research, there is usually an inkling on the part of the astronomer that a targeted instructional change in his or her classroom will benefit students and the goal is to determine what the learning benefits are. In action research, it would be rare for an astronomer to withhold an educational innovation believed to be beneficial from half of his or her students in an effort to establish a quasi-experimental comparison group. Nonetheless, action research provides an important mechanism for transforming the role of faculty from that of researchers who occasionally lecture to a perspective that recognizes teaching as a scholarly creative activity.

Example Action Research Projects

AR Project #1: *Can we quickly determine students' pre-course knowledge without using an extensive pre-test? This is important for modifying the pace of instruction and creating effective collaborative working groups. The approach was to survey students to find out how they rated their level of understanding of seven specific astronomy concepts both before and after instruction (pre-test/post-test strategy). The results were then matched to student performance on a 21-item multiple-choice test.*

Comparing pre-test to post-test gains we found: (1) there were statistically significant student gains on students' self-report of knowledge ($self_{pre}=2.36$ to $self_{post} = 3.71$ on a scale of 1 to 5) implying that the five-level self-report survey is sensitive enough to measure perceived gains in knowledge; (2) there were significant student gains on multiple-choice items ($MC_{pre}=50\%$ to $MC_{post}=70\%$) implying that learning did occur; (3) there was a reasonably high correlation between self-report and exam performance ($r_{pre}=.46$ and $r_{pos}=.39$ where $r=0$ is no correlation and $r=1$ is perfect correlation); and (4) males self-report slightly higher than females, but demonstrate no difference in performance. This analysis suggests that self-report gains are representative of actual student gains on multiple-choice scores and that students can accurately recognize and report their knowledge levels. It appears that, within the context of this class, *two-minute self-report surveys can be substituted for conventional 20-minute pre-test exams to estimate students' initial knowledge state.*

Example Action Research Projects (continued)

AR Project #2: Does required e-mail contact between students and faculty improve instructor-availability ratings on faculty evaluation forms? The approach was to award points to students for emailing the instructor twice during the semester. The students were encouraged to use the opportunity to initiate a meaningful interaction but understood that points were awarded irrespective of the content of the message—the instructor recorded in a personal journal the perceived meaningfulness of each electronic interaction. The instructor replied to all messages and, where email was not the appropriate medium to hold the discussion, followed up with a phone call.

In the semester prior to implementing this strategy, the instructor received 15 email messages from students and received an instructor availability rating of $\underline{x}$ =1.83 (where 1=good and 5=poor). With the new strategy in place, the instructor received 157 email messages, 149 of which were judged to be meaningful. In addition, many students, who likely would not have done so without some encouragement, continued to email the instructor. Surprising, however, the overall rating of instructor availability remained essentially constant at 1.88. These results suggest that, in this context, student ratings of instructor availability are not impacted by email communication even though the instructor's perception was an overall increase in meaningful interactions with students.

Example Action Research Projects (continued)

AR Project #3: *How do student writing skills correlate with exam performance? Students in this course are required to complete three one-page writing assignments, each graded for content, grammar, and style. The assigned topics are not highly technical and encourage creativity. For example, one writing prompt is as follows: "Since the time of Copernicus, we have known that the Earth goes around the Sun and yet newspapers still report the times that the Sun rises and sets, suggesting that the Sun goes around the Earth. Is it wrong to use a model that is inherently flawed?" This task is very different than the homework and exams, which focus much more on technical knowledge. We were interested in how well student performance on the writing task was related to their performance on the multiple-choice exams.*

We examined the correlation between each student's final exam score and her/his average writing score. A correlation analysis, easily performed on most spreadsheet programs, yielded a correlation coefficient of $r=0.50$. To get a better sense of the meaning of this number, we compared students' final exam scores with their scores on chapter tests and weekly homework. The correlation with the chapter tests was higher at 0.76. The correlation between final exam scores and homework scores was only 0.45, which is slightly lower than the correlation with writing assignments. The data contradicted our initial belief that, based on content similarity, the final exam scores would be more highly correlated to the homework than the writing assignments. This suggests a more integral connection between writing and test performance than we had anticipated—a connection that we wanted to take into account in any future course revisions.

Example Action Research Projects *(continued)*

AR Project #4: *What is working well and not working so well in implementing collaborative learning groups in the large lecture course? In the fall of 1997, we made major course revisions to our astronomy course. The goal was to increase student participation and attendance by incorporating frequent small group discussion activities into the lecture environment. We developed a series of sixteen mini-labs that students completed working in groups of four. Each activity required between 20 and 50 minutes to complete, with each student receiving the group score. Both quizzes and examinations contained a group component.*

To evaluate the implementation of this innovative approach, we enlisted 10 faculty from across campus to audio-tape exploratory focus group discussions with groups of up to 20 students. A survey of 10 hours of audio-tape revealed that: (1.) students enjoy the alternation between activity and lecture; (2.) students report learning from each other; (3.) students would prefer to have a more detailed reading list than was initially provided; (4.) the structure of the exam needed to be more clearly defined; and (5.) students would like to have more specific roles and responsibilities in their collaborative learning groups. The results of these interviews were reported to the class as a whole and, where feasible, changes were implemented to address students' concerns. End-of-course surveys indicated both that some of the students' concerns were addressed by our mid-course corrections and that the students appreciated the process. Students informally commented that this experience demonstrated that the instructors cared about students' learning.

Wrapping It Up: *Where Do We Go From Here?*

By now, we hope that you have gotten a feel for some of the basics of AER. Our intent was to provide a first-steps introduction, with astronomy-specific examples wherever possible, so that you might know where to begin this new adventure. To be sure, we haven't covered everything—there are already lots of books out there (we've referred to many!) that are more comprehensive than this one. These resources can be immensely valuable, and we encourage you to seek them out as you delve into AER and come across questions that this book doesn't answer.

AER is a nascent and vibrant field, and there's plenty of room to grow. Indeed, there are so many rich avenues for research available, we are pretty sure that there are more questions to pursue than any of us already doing some AER have time to research—so please join us! By performing careful and systematic investigations of relevant teaching and learning issues, we can help improve both the quantity and quality of astronomy education for the hundreds of thousands of students who take our classes every year.

References

Adams, J. P., & Slater, T. F. (1998). Using action research to bring the large lecture course down to size. *Journal of College Science Teaching, 28,* 87-90.

Anderson, L. W., & Bourke, S. F. (2000). *Assessing affective characteristics in the schools* (2nd ed.). Mahwah, NJ: Lawrence Erlbaum Associates.

Angelo, T. A., & Cross, K. P. (1993). *Classroom assessment techniques: A handbook for college teachers* (2nd ed.). San Francisco: Jossey-Bass, Inc.

Bailey, J. M., & Slater, T. F. (2003). A review of astronomy education research. *Astronomy Education Review, 2*(2), 20-45. Retrieved from http://aer.noao.edu/cgi-bin/article.pl?id=63

Bailey, J. M., & Slater, T. F. (2005). Finding the forest amid the trees: Tools for evaluating astronomy education and public outreach projects. *Astronomy Education Review, 3*(3), 47-60. Retrieved from http://aer.noao.edu/cgi-bin/article.pl?id=120

Bailey, J. M., & Slater, T. F. (2005). Resource letter AER-1: Astronomy education research. *American Journal of Physics, 73*(8), 677-685.

Baker, D. (Personal communication, March 31, 2008).

Bodner, G. M., & Orgill, M. (Eds.). (2007). *Theoretical frameworks for research in chemistry/science education.* Upper Saddle River, NJ: Pearson/Prentice Hall.

Bransford, J. D., Brown, A. L., & Cocking, R. R. (Eds.). (2000). *How people learn: Brain, mind, experience, and school: Expanded edition*. Washington, DC: National Academies Press.

Brogt, E., Dokter, E. F., & Antonellis, J. (2007a). Regulations and ethical considerations for astronomy education research. *Astronomy Education Review*, *6*(1), 43-49. Retrieved from http://aer.noao.edu/cgi-bin/article.pl?id=242

Brogt, E., Dokter, E. F., Antonellis, J., & Buxner, S. (2007b). Regulations and ethical considerations for astronomy education research II: Resources and worked examples. *Astronomy Education Review*, *6*(2), 99-110. Retrieved from http://aer.noao.edu/cgi-bin/article.pl?id=257

Brogt, E., Sabers, D., Prather, E. E., Deming, G. L., Hufnagel, B., & Slater, T. F. (2007). Analysis of the Astronomy Diagnostic Test. *Astronomy Education Review*, *6*(1), 25-42. Retrieved from http://aer.noao.edu/cgi-bin/article.pl?id=239

Brown, A. L. (1975). The development of memory: Knowing, knowing about knowing, and knowing how to know. *Advances in Child Development and Behavior, 10*, 103-152.

Brown, A. L. (1994). The advancement of learning. *Educational Researcher, 23*(8), 4-12.

Burns, M. (1993). *Mathematics: Assessing understanding*. White Plains, NY: Cuisenaire Company of America.

Campbell, D. T., & Stanley, J. C. (1966). *Experimental and quasi-experimental designs for research*. Chicago: Rand McNally.

Chickering, A., & Gamson, Z. (1987). Seven principles for good practice. *AAHE Bulletin, 39*, ED 282 491, 3–7. Retrieved from http://www.aahea.org/bulletins/articles/sevenprinciples1987.htm

Corbin, J. M., & Strauss, A. (2007). *Basics of qualitative research: Techniques and procedures for developing grounded theory* (3rd ed.). Thousand Oaks, CA: SAGE Publications, Inc.

Creswell, J. W. (1998). *Qualitative Inquiry and Research Design: Choosing Among Five Traditions*. Thousand Oaks, CA: Sage Publications, Inc.

Creswell, J. W. (2007). *Educational research: Planning, conducting, and evaluating quantitative and qualitative research* (3rd ed.). Upper Saddle River, NJ: Prentice Hall.

Erickson, F. (1986). Qualitative research on teaching. In M. C. Wittrock (Ed.), *Handbook for research on teaching* (3rd ed., pp. 119-161). New York: Macmillan.

Erlwanger, S. H. (1973). Benny's conception of rules and answers in IPI mathematics. *Journal of Children's Mathematical Behavior, 1*(2), 7-26.

Felder, R., Woods, D., Stice, J., & Rugarcia, A. (2000). The future of engineering education: II. Teaching methods that work. *Chemical Engineering Education, 34*(1), 26–39.

Francis, G. E., Adams, J. P., & Noonan, E. J. (1998). Do they stay fixed? *The Physics Teacher, 36*(8), 488- 491.

Glaser, B.G. (1978). *Theoretical sensitivity.* Mill Valley, CA: Sociology Press.

González, N., Moll, L. C., & Amanti, C. (2005). *Funds of knowledge: Theorizing practices in households, communities, and classrooms.* Mahwah, NJ: Lawrence Erlbaum Associates.

Hake, R. R. (1998). Interactive-engagement versus traditional methods: A six-thousand-student survey of mechanics test data for introductory physics courses. *American Journal of Physics, 66*(1), 64-74.

Heath, S. B. (1987). *Ways with words: Language, life, and work in communities and classrooms.* Cambridge, UK: Cambridge University Press.

Jaeger, R. M. (1990). *Statistics: A spectator sport* (2nd ed.). Thousand Oaks, CA: Sage Publications, Inc.

Johnson, D., Johnson, R., & Smith, K. (1998). *Active learning: Cooperation in the college classroom* (2nd ed.). Edina, MN: Interaction Book Company.

Laws, P., Sokoloff, D., & Thornton, R. (1999). Promoting active learning using the results of physics education research. *UniServe Science News, 13,* 14-19.

McKeachie, W. (1972). Research on college teaching. *Educational Perspectives, 11*(2), 3–10.

Mills, G. E. (2006). *Action research: A guide for the teacher researcher* (3rd ed.). Upper Saddle River, NJ: Prentice Hall.

Minstrell, J. (1989). Teaching science for understanding. In L. B. Resnick & L. E. Klopfer (Eds.), *Toward the thinking curriculum: Current cognitive research* (pp. 129-149). Alexandria, VA: Association for Supervision and Curriculum Development.

The National Commission for the Protection of Human Subjects of Biomedical and Behavioral Research. (1979). *The Belmont report: Ethical principles and guidelines for the protection of human subjects of research.* Washington, DC: National Institutes of Health. Retrieved from http://ohsr.od.nih.gov/guidelines/belmont.html

Palinscar, A. S., & Brown, A. L. (1984). Reciprocal teaching of comprehension-fostering and comprehension-monitoring activities. *Cognition and Instruction, 1*, 117–175.

Perrenoud, P. (1991). Towards a pragmatic approach to formative evaluation. In P. Weston (Ed.), *Assessment of pupils' achievement: Motivation and school success* (pp. 77-101). Amsterdam: Swets and Zeitlinger.

Piaget, J. (1954). *The construction of reality in the child*. New York: Basic Books.

Piaget, J. (1967). *The child's conception of the world*. London: Routledge & Kegan.

Redish, E., Saul, J., & Steinberg, R. (1997). On the effectiveness of active-engagement microcomputer-based laboratories. *American Journal of Physics, 65*(1), 45-54.

Rosenshine, B., & Meister, C. (1994). Reciprocal teaching: A review of the research. *Review of Educational Research, 64*, 479–530.

Ruhl, K. L., Hughes, C. A., & Schloss, P. J. (1987). Using the pause procedure to enhance learning recall. *Teacher Education & Special Education, 10*, 14-18.

Shepard, L. A. (1997). *Measuring achievement: What does it mean to test for robust understanding?*. Princeton, NJ: Policy Information Center, Education Testing Service.

Shepard, L. A. (2001). The role of classroom assessment in teaching and learning. In V. Richardson (Ed.), *Handbook of research on teaching* (4th ed.) (pp. 1066-1101). Washington, DC: American Educational Research Association.

Springer, L., Stanne, M., & Donovan, S. (1999). Effects of small-group learning on undergraduates in science, mathematics, engineering and technology: A meta-analysis. *Review of Educational Research, 69*(1), 21–52.

Sternberg, R. J. (1983). Criteria for intellectual skills training. *Educational Researcher, 12*, 6-12.

Stipek, D. J. (1996). Motivation and instruction. In D. C. Berliner & R. C. Calfee (Eds.), *Handbook of educational psychology* (pp. 85-113). New York: Simon & Schuster Macmillan.

Strauss, A. L., & Corbin, J. M. (1996). *Basics of qualitative research: Grounded theory procedures and techniques.* Newbury Park, CA: Sage Publications.

Sutherland, T. E., & Bonwell, C. C. (1996). *Using active learning in college classes: A range of options for faculty.* San Francisco: Jossey-Bass, Inc.

Thalheimer, W., & Cook, S. (2002, August). How to calculate effect sizes from published research: A simplified methodology. Retrieved from http://www.work-learning.com/effect_sizes.htm

Thornton, R. K., & Sokoloff, D. R. (1998). Assessing student learning of Newton's Laws: The Force and Motion Conceptual Evaluation and the evaluation of active learning laboratory and lecture curricula. *American Journal of Physics, 66*(4), 338-352.

Tobin, K., & Ulerick, S. (1989). An interpretation of high school science teaching based on metaphors and beliefs for specific roles. *Paper presented at the annual meeting of the American Educational Research Association.* San Francisco, CA.

Towne, L., & Shavelson, R. J. (2002). *Scientific research in education.* Washington, DC: National Academies Press.

United States Holocaust Memorial Museum. (2008, May 20). The doctors trial: The medical case of the subsequent Nuremberg proceedings. *Holocaust Encyclopedia*. Retrieved from http://www.ushmm.org/wlc/article.php?lang=en&ModuleId= 10007035

Yackel, E., Cobb, P., & Wood, T. (1991). Small-group interactions as a source of learning opportunities in second-grade mathematics. *Journal for Research in Mathematics Education, 22*, 390-408.

Appendix A
Test Of Astronomy STandards (TOAST)

Introductory Astronomy Survey
TOAST 060608

*Answer all of the following questions on the bubble sheet provided using a #2 pencil. Make sure that your **Name and Student ID Number** are on the answer sheet and that all answers are recorded in the correct position. **Always select the BEST answer**.*

This is a voluntary survey to help us improve and focus this class. Your participation in no way impacts your grade. We ask for your name and ID only so that we can match your answers with those on future voluntary surveys. If you have questions, please raise your hand.

DO NOT START THE SURVEY UNTIL YOU ARE TOLD TO DO SO!

PLEASE DO NOT WRITE ON THIS SURVEY

Use the drawing below to answer the first <u>two</u> questions.

1. If you could see stars during the day, the drawing below shows what the sky would look like at *noon* on a given day. The Sun is at the highest point that it will reach on this day and is near the stars of the constellation Gemini. What is the name of the constellation that will be closest to the Sun at sunset on this day?

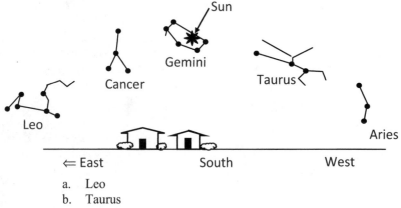

a. Leo
b. Taurus
c. Aries
d. Cancer
e. Gemini

2. This picture shows the position of the stars at *noon* on a certain day. How long would you have to wait to see Gemini at this same position *at midnight?*
   a. 12 hours
   b. 24 hours
   c. 6 months
   d. 1 year
   e. Gemini is never seen at this position at midnight.

3. You look to the eastern horizon as the Moon first rises and discover that it is in the new moon phase. Which picture shows what the moon will look like when it is at its high point in the sky, later that same day?

   a. A
   b. B
   c. C
   d. D
   e. E

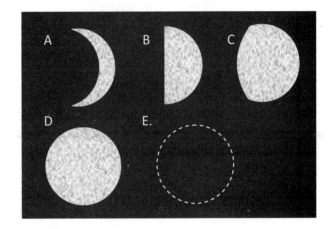

4. You are located in the continental United States on the first day of October. How will the position of the Sun at noon be different two weeks later?

   a. It will have moved toward the north.
   b. It will have moved to a position higher in the sky.
   c. It will stay in the same position.
   d. It will have moved to a position closer to the horizon.
   e. It will have moved toward the west.

5. Which sentence best describes why the Moon goes through phases?

   a. Earth's shadow falls on different parts of the Moon at different times.
   b. The Moon is somewhat flattened and disk-like. It appears more or less round depending on the precise angle from which we see it.
   c. Earth's clouds cover potions of the Moon resulting in the changing phases that we see.
   d. The sunlight reflected from Earth lights up the Moon. It is less effective when the Moon is lower in the sky than when it is higher in the sky.
   e. We see only part of the lit-up face of the Moon depending on its position relative to Earth and the Sun.

6. Imagine you see Mars rising in the east at 6:30 pm. Six hours later what direction would you face (look) to see Mars when it is highest in the sky?
   a. toward the north
   b. toward the south
   c. toward the east
   d. toward the west
   e. directly overhead

7. Imagine that Earth was upright with no tilt. How would this affect the seasons?
   a. We would no longer experience a difference between the seasons.
   b. We would still experience seasons, but the difference would be *less* noticeable.
   c. We would still experience seasons, but the difference would be *more* noticeable.
   d. We would continue to experience seasons in essentially the same way we do now.

8. How does the Sun produce the energy that heats our planet?
   a. The gases inside the Sun are burning and producing large amounts of energy.
   b. Gas inside the Sun heats up when compressed, giving off large amounts of energy.
   c. Heat trapped by magnetic fields in the Sun is released as energy.
   d. Hydrogen is combined into helium, giving off large amounts of energy.
   e. The core of the Sun has radioactive atoms that give off energy as they decay.

9. The Big Bang is best described as:
   a. The event that formed all matter and space from an infinitely small dot of energy.
   b. The event that formed all matter and scattered it into space.
   c. The event that scattered all matter and energy throughout space.
   d. The event that organized the current arrangement of planetary systems.

10. Which of the following ranks locations from <u>closest to Earth to farthest from Earth</u>?
    a. the Sun, the Moon, the edge of our solar system, the North Star, the edge of our galaxy
    b. the Sun, the North Star, the Moon, the edge of our galaxy, the edge of our solar system
    c. the Moon, the North Star, the Sun, the edge of our solar system, the edge of our galaxy
    d. the Moon, the Sun, the edge of our solar system, the North Star, the edge of our galaxy
    e. the North Star, the Moon, the Sun, the edge of our galaxy, the edge of our solar system

---

Consider the six different astronomical objects (A-F) shown below.

A. The Solar System

B. The Sun

C. Jupiter

D. Andromeda Galaxy

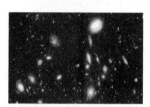

E. Galaxy Cluster

F. Nebula

11. Which of the following is the best ranking (from smallest to largest) for the size of the above objects?
    a. C<F<B<A<D<E
    b. E<D<F<A<B<C
    c. C<B<A<F<D<E
    d. F<C<B<A<D<E
    e. None of the above is correct.

---

12. Imagine that Earth's orbit were changed to be a perfect circle around the Sun so that the distance to the Sun never changed. How would this affect the seasons?
    a. We would not be able to notice a difference among seasons.
    b. The difference in the seasons would be *less* noticeable than it is now.
    c. The difference in the seasons would be *more* noticeable than it is now.
    d. We would experience seasons in the same way we do now.

13. What is a star?
    a. a ball of gas that reflects light from another energy source
    b. a bright point of light visible in Earth's atmosphere
    c. a hot ball of gas that produces energy by burning gases
    d. a hot ball of gas that produces energy by combining atoms into heavier atoms
    e. a hot ball of gas that produces energy by breaking apart atoms into lighter atoms

14. Which one property of a star will determine the rest of the characteristics of that star's life?
    a. brightness
    b. temperature
    c. color
    d. mass
    e. chemical makeup

15. Current evidence about how the universe is changing tells us that
    a. we are near the center of the universe.
    b. galaxies are expanding into empty space.
    c. groups of galaxies appear to move away from each other
    d. nearby galaxies are younger than distant galaxies.

16. Stars begin life as
    a. a piece off of a star or a planet.
    b. a white dwarf.
    c. matter in Earth's atmosphere.
    d. a black hole.
    e. a cloud of gas and dust.

17. When the Sun reaches the end of its life, what will happen to it?
    a. It will turn into a black hole.
    b. It will explode, destroying Earth.
    c. It will lost its outer layers, leaving its core behind.
    d. It will not die due to its mass.

18. If you were in a spacecraft near the Sun and began traveling to Pluto you might pass
    a. planets.
    b. stars.
    c. moons.
    d. two of these objects.
    e. all of these objects.

19. How did the system of planets orbiting the Sun form?
    a. The planets formed from the same materials as the Sun.
    b. The planets and the Sun formed at the time of the Big Bang.
    c. The planets were captured by the Sun's gravity.
    d. The planets formed from the fusion of hydrogen in their cores.

20. Which of the following would make you weigh half as much as you do right now?
    a. Take away half of the Earth's atmosphere.
    b. Double the distance between the Sun and the Earth.
    c. Make the Earth spin half as fast.
    d. Take away half of the Earth's mass.

21. Astronauts "float" around in the space shuttle as it orbits Earth because
    a. there is no gravity in space.
    b. they are falling in the same way as the Space Shuttle.
    c. they are above Earth's atmosphere.
    d. there is less gravity inside of the Space Shuttle.

22. Energy is released from atoms in the form of light when electrons
    a. are emitted by the atom.
    b. move from low energy levels to high energy levels.
    c. move from high energy levels to low energy levels.
    d. move in their orbit around the nucleus.

23. Which of the following would be true about comparing visible light and radio waves?
    a. The radio waves would have a lower energy and would travel slower than visible light.
    b. The visible light would have a shorter wavelength and a lower energy than radio waves.
    c. The radio waves would have a longer wavelength and would travel the same speed as visible light.
    d. The visible light would have a higher energy and would travel faster than radio waves.
    e. The radio waves would have a shorter wavelength and higher energy than visible light.

24. The atoms in the plastic of your chair were formed
    a. in our Sun.
    b. by a star existing prior to the formation of our Sun.
    c. at the instant of the Big Bang.
    d. approximately 100 million years ago.
    e. in a distant galaxy in a different part of the early universe.

Use the drawings below to answer the next two questions.

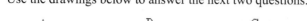

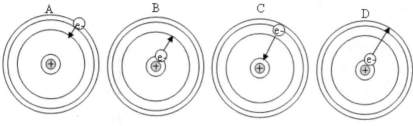

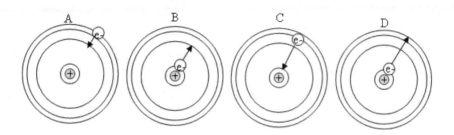

25. Which atom would be absorbing light with the greatest energy?
    a. A
    b. B
    c. C
    d. D

26. Which atom would emit light with the shortest wavelength?
    a. A
    b. B
    c. C
    d. D

27. The graphs below illustrate the energy output versus wavelength for three unknown objects A, B, and C. Which of the objects has the highest temperature?

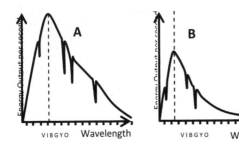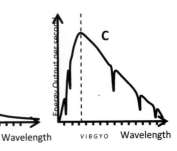

   a. A
   b. B
   c. C
   d. All three objects have the same temperature.
   e. The answer cannot be determined from this information.

ATTITUDES TOWARD SCIENCE INSTRUMENT

		Strongly Agree	Agree	Neither Agree nor Disagree	Disagree	Strongly Disagree
1.	When I hear the word "science," I have a feeling of dislike.	SA	A	N	D	SD
2.	I feel tense or upset when someone talks to me about science.	SA	A	N	D	SD
3.	It makes me nervous to even think about doing science.	SA	A	N	D	SD
4.	It scares me to have to take a science class.	SA	A	N	D	SD
5.	I have a good feeling toward science.	SA	A	N	D	SD
6.	Science is useful for solving the problems of everyday life.	SA	A	N	D	SD
7.	Most people should study some science.	SA	A	N	D	SD
8.	Science is helpful in understanding today's world.	SA	A	N	D	SD
9.	Science is of great importance to a country's development.	SA	A	N	D	SD
10.	It is important to know science in order to get a good job.	SA	A	N	D	SD
11.	I do not do very well in science.	SA	A	N	D	SD
12.	Science is easy for me.	SA	A	N	D	SD
13.	I usually understand what we are talking about in science.	SA	A	N	D	SD
14.	No matter how hard I try, I cannot understand science.	SA	A	N	D	SD
15.	I often think, "I cannot do this," when a science assignment seems hard.	SA	A	N	D	SD
16.	Science is something that I enjoy very much.	SA	A	N	D	SD
17.	I like the challenge of science assignments.	SA	A	N	D	SD
18.	It is important to me to understand the work I do in the science class.	SA	A	N	D	SD
19.	Science is one of my favorite subjects.	SA	A	N	D	SD
20.	I have a real desire to learn science.	SA	A	N	D	SD

Appendix C
Research Involving Human Subjects (IRB)

C learly, a major difference between astronomical research and AER is the involvement of people as the subject of the research question. This adds a layer of complexity to a study that has to be addressed before any data can be collected. The intention of this appendix is to give you an overview of some of the considerations and procedures. Oversight of human subjects research is done at the institutional level based upon federal regulations; thus, each researcher will have to investigate their own institution's policies and procedures and, naturally, we can't cover it all here.

Background and Ethical Considerations

In the mid 20th century, information was brought to light about a group of medical research studies that were (later) determined to be unethical. Experimentation on prisoners within the Nazi concentration camps of World War II has been well-documented and was a factor in the trials of many war criminals (United States Holocaust Memorial Museum, 2008). Another example, known as the Tuskegee Syphilis Study but run by the U.S. Department of Public Health, was a 40-year study that enticed nearly 400 African-American males to accept "free treatments" for syphilis, when in fact they received placebos. Eventually, 128 of the subjects were dead from the disease or related complications.

From these and other similar incidents came a number of international agreements and federal regulations about the use of experimentation with human beings. While medical research was long the basis for such regulations, social and behavioral sciences (such as education and psychology) have increasingly been governed by similar procedures. The Nuremberg Code, adopted as part of the Nazi war crime trials of 1945–1946, and the Belmont Report (The National Commission for the Protection of Human Subjects of Biomedical and Behavioral Research, 1979), are the primary documents that guide the regulations.

The most basic tenants of human subjects research include the following three ideas.

a. **"*Respect for Persons*"** – Persons are individual and autonomous, and as such are afforded a respect in regards to their ability to make decisions about what they are and are not willing to do. Certain populations who may have diminished capacity are classified as "vulnerable" and are protected by additional regulations.

b. **"*Beneficence*"** – Researchers must not only protect their research subjects from harm, they are also obligated to maximize possible benefits.

c. **"*Justice*"** – Persons should be treated equally, both in general and insofar as the details of the particular research study are concerned.

These tenets have led to the requirement that potential research projects be carefully planned and reviewed in regards to the recruitment and selection of potential subjects and the appraisal of potential risks and benefits. Furthermore, potential subjects should provide informed consent to participate in the project.

Institutional Support and Training

Each institution should have an office that works with researchers involved in studies of human subjects. While the exact name might be different (e.g., Office for the Protection of Research Subjects or Human Subjects Protection Program), each is basically an office in support of the *Institutional Review Board*, or IRB. The IRB is a committee of researchers and administrators who oversee all relevant research. These offices are often housed under your institution's Vice President for Research or similar office.

Many institutions require some level of training before you can undertake research with human subjects. Training may consist of, for example, a workshop, a self-paced study course and exam, or a web-based tutorial. Many organizations have subscribed to the University of

Miami's Collaborative Institutional Training Initiative (CITI). This web-based tutorial system allows researchers from any affiliated organization to go through the same process and exams (perhaps with an additional module for institution-specific policies) as a researcher at a different institution. This is particularly helpful when you have collaborators from different institutions, as your institution can feel confident about the training that person has had.

IRB Proposals

An IRB will likely require some level of proposal for any research project involving human subjects. Because each institution's IRB handles things their own way (though under guidance from the federal regulations), it is impossible for us to provide details about how you will maneuver through your own system. However, there are likely to be universal issues that will need to be discussed in some way. The table below provides a list of some of the most common topics.

Common Topics That Will Need to be Reviewed by the IRB

- *Participant selection and, if applicable, exclusion*
- *Participant recruitment*
- *Obtaining informed consent*
- *Anticipated risks*
- *Anticipated benefits*
- *Costs of participation*
- *Compensation for participation*
- *Timeline*
- *Process of data collection*
- *Description of data to be collected*
- *Where and how data will be securely held, and eventually destroyed*

Carefully consider these issues before you begin to work on your IRB proposal. Of course, there will probably be additional information required on the proposal. If you are unsure about how to answer something, contact your IRB office or, if you have one, a department or college representative. Most are happy to help. You might also ask your IRB office or an experienced colleague for an example to see what a successful proposal looks like. Upon review, the IRB may ask for modifications to your proposal. In our experience, it is best to do exactly what they ask for even if it may not make sense to you. Trust us, they're the experts!

> **EXIT**
>
> More information about the IRB as it relates to AER can be found in two recent articles in *Astronomy Education Review* (Brogt, Dokter, & Antonellis, 2007a; Brogt, Dokter, Antonellis, & Buxner, 2007b).

You'll likely also have to do annual reports or reviews if your project takes more than a year. These may be as simple as a checklist or as detailed as a report, and they may be due as much as 60 days in advance of your anniversary date. Find out your institution's procedures and follow them carefully. If you miss a deadline, you are forbidden to continue to collect or analyze your data until the process has been completed.

Finally, a reminder and a piece of advice. Despite any rumors to the contrary, the IRB and its members really do want to support you in pursuing your research topics and are not out to make you kill lots of trees with their vast amounts of paperwork. They are tasked with making sure that research is done in an ethical and responsible manner and follows all legal regulations where relevant. The best way to be successful is to ask questions, provide as much detail and information as you can in your proposal and if asked later, and keep your participants in the forefront of your mind. If you do these things, it won't seem such an arduous task.

FINE PRINT: This book cannot be considered as legal advice. Always contact YOUR institution's IRB for specific policies, procedures, and assistance.

About the Authors

Janelle M. Bailey, Ph.D., is an Assistant Professor of Science Education in the Department of Curriculum & Instruction at the University of Nevada, Las Vegas. Her research interests include the teaching and learning of science and the effectiveness of professional development for science teachers. She teaches courses in science education, including methods and research courses, for both undergraduate and graduate students. She is the past Chair of the American Association of Physics Teachers' Space Science and Astronomy Committee. Dr. Bailey earned her B.A. in Astrophysics from Agnes Scott College and her M.Ed. in Science Education from the University of Georgia. Her Ph.D. is from the University of Arizona's Department of Teaching and Teacher Education, where she studied undergraduates' conceptual understanding of stars and star properties.

Stephanie J. Slater, Ph.D., is an Assistant Professor at the University of Wyoming in the Department of Elementary and Early Childhood Education and a Research Scientist in the Department of Physics and Astronomy. She is responsible for teaching science methods courses that are based on inquiry learning and for graduate research methods courses related to the teaching of astronomy. She is the Director of Research for the *Center for Astronomy & Physics Education Research* (CAPER) Team, where her research focuses on student conceptual understanding as influenced by students' spatial reasoning abilities and cognitive load, and inquiry-based curriculum development with a particular emphasis on preparing future teachers. She earned a M.S. in Interdisciplinary Science–Science Education from Montana State University and a B.S. in Biology and Mathematics from Harding University. Her Ph.D. is from the University of Arizona in the Department of Teaching and Teacher Education, where she studied the educational nature of undergraduate research experiences for science majors in science career pathways.

Timothy F. Slater, Ph.D., is a Professor at the University of Wyoming where he holds the Wyoming Excellence in Higher Education Endowed Chair for Science Education. He leads the *Center for Astronomy & Physics Education Research* (CAPER) Team, where his research focuses on student conceptual understanding in formal and informal learning environments, inquiry-based curriculum development, and authentic assessment strategies, with a particular emphasis on non-science majors and pre-service teachers. Prior to joining the faculty at the University of Wyoming, Dr. Slater was a tenured professor in the Astronomy Department at the University of Arizona where he built the first Ph.D. program in astronomy education research. Professor Slater earned his Ph.D. at the University of South Carolina in geological sciences and his M.S. from Clemson University in Astronomy. He holds two bachelors' degrees from Kansas State University, one in Science Education and one in Physical Science. Professor Slater has served as the elected education officer for the American Astronomical Society, an elected member of the Board of Directors for the Astronomical Society of the Pacific, the Board of the National Science Teachers Association, an elected councilor at large for the Society of College Science Teachers, a member of the founding editorial board of the *Astronomy Education Review*, and multiple terms as chairman of the Astronomy Education Committee of the American Association of Physics Teachers. He is an author on nearly 100 refereed articles and eight books, is the winner of numerous awards, and is frequently an invited speaker on improving teaching of science through educational research and improving teacher education.